Toward Scientific Medicine

O. S. Miettinen

Toward Scientific Medicine

O. S. Miettinen
Joint Departments of Epidemiology & Biostatistics
  and Occupational Health
McGill University
Montreal, Québec
Canada;
and
Department of Medicine
Weill Medical College of Cornell University
New York, NY, USA

ISBN 978-3-319-34633-5   ISBN 978-3-319-01671-9 (eBook)
DOI 10.1007/978-3-319-01671-9
Springer Cham Heidelberg New York Dordrecht London

Printed on acid-free paper

Springer is part of Springer Science+Business Media (www.springer.com)

Foreword

While medicine has splintered into a multitude of separate disciplines of practice, medical academia too has become very diverse in its particular disciplines of scholarship. This diversity in medical academia has created a need for one of the disciplines to be focused on the 'big picture' of medicine, on the scholarly underpinnings of medicine at large, especially the general theory of medicine itself and also the theory of directly practice-relevant medical research. Even though still generally neglected, the scholarly underpinnings of medicine at large and of the research to advance the knowledge base of medicine should be areas of central concern in academic schools of medicine, for both the advancement and teaching of these.

As an outstanding exception, these matters have been of central concern, and for a very long time already, for Professor Miettinen. In this book – the very first of its kind by anyone – he first delineates the "daunting" needs for knowledge in medicine, and he then explores, critically, the "knowledge culture" that now prevails in and around medicine, focusing on examples from textbooks, journals, and educational programs. He makes plain how vacuous in substance the knowledge base of medicine still is and how wanting even in general academic standards of quality these sources of medical knowledge still are prone to be, almost implying that 'medical academia' remains a contradiction in terms. And from this novel, and to most medical academics presumably surprising, diagnosis about the status quo he proceeds to presentation of what he sees to be the needed remedies. As the diagnosis reveals a severely anomalous state of the prevailing culture of medical academia – he points out that even the 'knowledge-based medicine' mantra is absent while 'evidence-based medicine' now is all around – the remedies he posits are correspondingly wide-ranging and radical, for both the knowledge culture of and surrounding medicine and the research to advance the knowledge base of medicine.

As I reflect on all of this, two thoughts are foremost in my mind. First, the range, depth, and originality of medical scholarship that are manifest in this book keep me wondering whether anyone other than Miettinen could have written a treatise like this, and I just haven't been able to think of anyone else as being anywhere close to measuring up. For all I can see, this book, timely and very important, had to be written by Miettinen.

And second, I am mindful of how medical academia in the past has initially dealt with ideas about necessary corrective innovations in medicine itself, before ultimately accepting and implementing them. In some cases the innovator has been driven out of town and also out of his mind (Semmelweis, inter alia), and so I wonder about the academia's initial reception of Miettinen's ideas about the need for wide-ranging, radical innovations in and around medicine, most of which, at least, deserve to be accepted and implemented.

A propos, Miettinen commonly quotes Francis Bacon's counsel about reading (the earlier, more succinct version of this): "Read not to contradict, nor to believe, but to weigh and consider." This indeed is the way this book of his own also should be read, and I add that his ideas about scientific medicine and medical science deserve particularly careful weighing and considering by leading academics concerned with medicine. The way in which, the extent to which, and the pace with which this will be done will be telling measures of how academic our particular, proud province of academia truly is. I hope it'll adopt a posture of humility and commitment to early implementation of the needed innovations.

Professor and Director   Johann Steurer, MD
Horten Center for Patient-oriented Research
   and Knowledge Transfer
Faculty of Medicine, University of Zurich
Zurich, Switzerland

Preface

In the 1990s, there was a reunion of my medical school class such that two of the members – both professors abroad – had been invited to give a talk to stimulate thought by the rest of us.

The first speaker, a 'basic' scientist, focused on a book that had been published 20 years earlier. This book had specified various innovations that the 'basic' sciences serving medicine will bring to the practice of medicine in the next two decades. The point of this talk was that none of those predicted innovations had materialized.

As the second speaker, I called attention to the evident fact that, after 6 years of shared studies we diverged to different lines of knowledge-dependent work, and that practically none of the requisite knowledge for these varied lines of work did we learn in the 6 years of shared study of "medicine."

Various colleagues who listened to us speakers subsequently expressed agreement with the disturbing implications of what had been said, but this they did only furtively, treating those implications as though they were closely guarded secrets of the medical guild. Open speaking about them was seen to be detrimental to one's career and was shunned on this basis.

Decades earlier, as a medical student, I was invited to join a group of physicians working on laboratory-based medical research (cardiologic), and I did, but the remoteness of this research from the practice of medicine and the expected deference in it to incomprehensible inputs from a statistician in another institute whetted my interest in more directly 'applied' medical research and a concern to achieve true and comprehensive understanding of the theoretical underpinnings of this much more heavily statistics-dependent genre of medical research. I was advised to go and study epidemiology and biostatistics, which I did. After 4 years of these studies I accepted an invitation to become an Assistant Professor of both of these disciplines (at Harvard), which was soon followed by promotions to the ranks of Associate Professor and Professor of these two disciplines.

Throughout the ensuing four decades of (full) professorship in these disciplines I've continued my quest – monomaniacal and quite lonely – to truly understand the nature of medicine's requisite knowledge base and the research to advance this knowledge. Even though this understanding of mine still is quite incomplete, I now feel it's time that I come out and share with others what my journey in medical academia, which began in 1956, has thus far led me to understand about the scholarly underpinnings of the vast industry of professional healthcare.

A physician's practice of healthcare can be in the service of a population, as a population, or it serves individuals one at a time. Correspondingly, medicine comprises community medicine or epidemiology and clinical medicine, the latter by far the dominant segment. In this book the focus is on clinical medicine, and the term 'medicine' is used in this somewhat limited meaning of it.

Two, very different, conceptions of *scientific medicine* came to eminence in the twentieth century, both of them still very influential, but my concept of scientific medicine is very different from both of those. To me, medicine is scientific to the extent that its theoretical framework is rational and its knowledge base derives from medical science – from research to advance the knowledge-base of rational medicine, from this quintessentially 'applied' genre of medical research. This research encompasses original research, producing new evidence to advance the knowledge-base of rational medicine, and derivative (review-type) research, collating and synthesizing evidence from such original research. The evidence on a given topic serves to advance the knowledge about it through reflection and public discourse in the scientific community concerned with the topic.

This book is about this rationalist conception of scientific medicine. It comprises three parts. Part I first outlines the nature and requisite form of codification of the knowledge-base of rational medicine – of gnosis (dia-, etio-, and prognosis) in it – and then, against this backdrop, it depicts the present state of directly practice-relevant knowledge and of the research to advance this knowledge. Consequent to what thus emerges, Part II presents the innovations that now are needed in the knowledge culture of medicine. Part III is mainly an introduction to the theory – concepts and principles, cum terminology (English) – of the research to advance the knowledge-base of genuinely scientific medicine, but to an extent it also is an introduction to the development of practice-guiding expert systems, initially pre-scientific but gradually also scientific. This Part III also is, in many ways, critical of the status quo, while mainly outlining the nature of the requisite research for genuinely scientific medicine.

This book is directed to two very different types of reader: academic leaders of medicine for one, and researchers concerned to advance the knowledge-base of medicine for another. For the former, all of Part II is very relevant, and so is the Epilogue, while perhaps only the chapter Abstracts in Part I and Part III are. For the researchers, Part I and Part II constitute an introduction to what ultimately matters to them, namely Part III. Relevant background readings for the researchers also are, I suggest, the books listed below, given that quintessentially applied clinical research is meta-epidemiological in nature, as it too addresses rates, though not rather non-specific rates, per se, as does epidemiological research, but very specific rates so as to learn about probabilities (gnostic).

My critical judgments about the status quo and corresponding remedial propositions concerning the knowledge culture surrounding and characterizing medicine likely will not be received with pleasure by many academic leaders of medicine, just as conveyance of diagnoses about serious illnesses and needs for radical interventions challenge patients' acceptance of such unwelcome truths about themselves. But receptivity to previously unrecognized troublesome truths is a prerequisite for remedying them in medicine itself just as in its clients, and the ultimate, constructive aim here is to prod the leaders, and the researchers too, to become agents of the much-needed innovations of the knowledge culture around and in medicine and, then, of actual progress toward genuinely scientific medicine.

In the balance hangs not only the somatic well-being and survival of people but people's material well-being as well, for from progress in medical academia flows improved quality of the practice of medicine, and from improved quality of the practice of medicine flows not only improved health and enhanced longevity of people but also improvement in the societal affordability of the healthcare people need.

Suggested introductory readings for researchers:

1. Miettinen OS. *Epidemiological Research: Terms and Concepts*. Dordrecht: Springer, 2011.

2. Miettinen OS, Karp I. *Epidemiological Research: An Introduction*. Dordrecht: Springer, 1212.

3. Miettinen OS. *Up from Clinical Epidemiology & EBM*. Dordrecht: Springer, 2011.

Acknowledgments

A decade ago, when contemplating the possibility of writing this book, I was being held back by a discouraging question: Does anyone of eminence in medicine really care to see the Flexnerian and Sackettian conceptions of scientific medicine (Sect. 4.1) give way to the idea that, actually, medicine should be scientific in the meaning of being not thinking-based (Flexner) nor evidence-based (Sackett) but knowledge-based – scientifically knowledge-based, and this in a rational, logically tenable theoretical framework? (Cf. Preface.)

Then to the fore came *Johann Steurer* from the Horten Center for Patient-oriented Research and Knowledge Transfer (of the Faculty of Medicine of the University of Zurich). As was evident even from the very name of the center he had formed, he already was pioneering that which I was only dreaming about. He already was fostering the practice of knowledge-based medicine, with the knowledge for it derived from research expressly designed to be 'applied' in this quintessentially pragmatic meaning of this word.

Much more than this initial, merely passive encouragement was going to be Steurer's influence on the development of this piece of work. Through the active communication and collaboration the two of us have had for a decade by now, he was, by far, the principal extrinsic impetus for growth in my sense of purpose and in my confidence about the basic tenets underpinning this project, and so I recently got to feel ready to take up the writing. And in the end, for good measure, Steurer enthusiastically, and with characteristic grace and humility, agreed to write the Foreword to this text. So, to whatever extent this oeuvre actually will serve the developmental purposes of Asklēpiós, the credit will largely belong to Steurer.

A source of inspiration also was the keen and sustained interest of my junior colleague *Igor Karp*, and his abundant, Socratic-type questions led to some notable improvements in the text. He was the junior Steurer in stimulating the genesis of the guideposts that here are posited for the path to genuinely scientific medicine.

Another former student of mine, *Malcolm Maclure*, read a late draft of this text and offered constructive comments.

In a different vein, a critical contribution was the modern 'word processing' that had to follow my pre-modern, pencil-and-paper type word production. This contribution was provided with wonderful dedication and expertise by *Dolores Coleto*.

On Medicine's Bonds with Science

"The dream of reason did not take power into account. . . . In America, no one group has held so dominant a position in this new world of rationality and power as has the medical profession. . . . The medical profession has had an exceptionally persuasive claim to authority. Unlike law and clergy, it enjoys close bonds with modern science, and at least for most of the last century, scientific knowledge has had a privileged status in the hierarchy of belief. . . . Its practitioners . . . serve as intermediaries between science and private experience, interpreting personal experience in the abstract language of scientific knowledge. . . ."

Starr P. *The Social Transformation of American Medicine. The Rise of a Sovereign Profession and the Making of a Vast Industry.* New York: Basic Books, Inc., Publishers, 1982. (pp. 3–4.)

Contents

Part I
The Knowledge-Base of Medicine at Present

Chapter 1
The Daunting Needs for Knowledge

Contents

1.0 Abstract

When a physician is practicing in the role of an actual *doctor*, his/her proximal aim in respect to a given client is to teach them about their health (L. *doctor*, 'teacher').

To this teaching end the physician needs to achieve – by deployment of general medical knowledge – knowing about health specific to the particular client at the particular time: *diagnosis* about the client's existing state of health as to the presence/absence of some particular illness(es); possibly *etiognosis* about antecedents of the client's illness as possible causes of this; and, regardless, *prognosis* about the client's future course of health, including as to the way this would depend on the choice of lifestyle and/or intervention (preventive or therapeutic).

For each of these purposes the requisite knowledge-base – probabilistic – is dauntingly complex, even for each of the already highly differentiated disciplines of principally knowledge-dependent, rather than skills-based, disciplines of modern medicine.

O. S. Miettinen, *Toward Scientific Medicine*, DOI 10.1007/978-3-319-01671-9_1,
© Springer International Publishing Switzerland 2014

1.1   Doctors' Dependence on Knowledge

In a given encounter with a client a physician usually deals with someone who in this encounter is his/her *patient*: a person who is suffering from an existing case of sickness (subjective and/or objective) and is looking for advice about dealing with this sickness (L. *pati*, 'suffer'). Otherwise the client in a given encounter is not a patient: (s)he may be seeking certification of freedom from some particular illness(es) (for life insurance, say) or potential 'early' detection of a particular illness (for its early, latent-stage treatment, if detected); or (s)he may be looking for advise on maintenance of good health.

At issue here is physician in the particular meaning and role of (medical) *doctor*: physician who teaches the client about their own health (L. *docere*, 'teach'). This teaching naturally is to be predicated on the requisite knowing – *esoteric knowing about the client's health*. This knowing is esoteric because it is attainable only by doctors, as they alone possess the requisite *general (abstract) medical knowledge* to decide what set of (particularistic) facts on the case to assemble and how to translate this set of ad-hoc facts into knowing about a hidden aspect of a person's health.

A doctor is, thus, critically dependent on general medical knowledge for the attainment of the particularistic knowing needed about, and by, the client. The client deploys this knowing in the framework of his/her personal *values*, in his/her *decisions* about health-related choices of action (testings, interventions, lifestyles).

1.2   The Knowledge Needed for Diagnosis

When a person with a complaint about some sickness presents to a physician, the first thing the physician is to know is, whether a presentation like this is in his/her ken to deal with, given his/her particular discipline ('specialty') of medicine. Even a 'general practitioner,' so called, may need to make an immediate referral to another discipline of medicine.

When taking on a case of sickness, the clinician first sets out to learn what is causing it, what illness (somatic anomaly) or what non-illness circumstance (medication use, say). In this pursuit, the first thing (s)he needs to know is: given the type of case presentation, what therefore are all the items of information (s)he needs to ascertain from 'history' (incl. the present) and physical examination, so as to establish the *clinical profile* of the case, representing the realizations of the full set of relevant clinical-type *diagnostic indicators* before any laboratory data.

Having established the clinical profile of the case, the physician proceeds to consideration of the corresponding *clinical diagnoses*. In this the first need is to know the complete set of possible causes of the type of sickness that is at issue, the *differential-diagnostic set* of possible causes. Included in this set may be some non-illness elements (cf. above).

Given the differential-diagnostic set, the physician needs to know, first, which ones of the elements in it, if any, are of urgent concern for management action, if indeed present. And insofar as there are possibilities of urgent concern, (s)he needs to know, first, the respective clinical (pre-lab) *probabilities* of their presence. Then, insofar as these probabilities are not extreme enough for action predicated on the presence of a particular one of these conditions, or inaction predicated on the absence of each of them, (s)he needs to know what non-clinical – referral-requiring, laboratory-based – *tests* would serve to make the probabilities for the urgent possibilities sufficiently extreme. (Electrocardiography and ultrasonography, for example, can in certain situations be viewed as being clinical tests, but CT imaging, for example, cannot.)

If all the urgent possibilities get to be (practically) ruled out, or there are none of these to begin with, the doctor needs to know the probabilities for each of the non-urgent possibilities and, if need be, what testing is called for to achieve a (practical) rule-in diagnosis about one of them (and thereby, generally, rule-out diagnosis about each of the others).

Insofar as referral-requiring testing is invoked, the doctor needs to know, in principle, the diagnostic probabilities conditional on each of the various possible *post-test profiles* of the case, representing enlarged counterparts of the pre-test profile, and in particular the probabilities conditional on whatever turns out to be the test result (commonly multidimensional, incl. from imaging alone) as an addition to the pre-test profile. When the testing is directed to a particular possibility (its ruling-in or ruling-out), the post-test probability of particular concern is that for the possibility at issue in the testing.

Rule-in diagnosis about one of the possible causes of the case of sickness at issue is generally taken, a priori, to be tantamount to ruling-out all of the other possibilities. With only illnesses in the differential-diagnostic set, diagnosis can be taken to be more descriptive than actually causal about the case in question. For, when an illness that is a possible cause of the case of sickness is deemed to actually be present, it is taken to be *the* cause of the sickness. This reasoning does not apply to any non-illness element in the differential-diagnostic set, which really are objects of etiognosis (Sect. 1.3) rather than diagnosis.

Relevant to all of this is the doctor's possession of the proper *concept of diagnosis*. The beginning in this is making the necessary distinction between the possible presence of a particular illness and the doctor's knowing about the presence/absence of this illness. Diagnosis in the context of a case of sickness, properly understood, is *knowing* about the presence/absence of a particular illness potentially explanatory of it (Gr. *dia* -, 'through,' 'apart'; *gignōskein*, 'perceive,' 'know'; *gnosis*, 'esoteric knowledge'). It is a doctor's first-hand knowing – gnosis – about this, distinct from the corresponding second-hand knowing by the client (consequent to 'doctoring' in the word's proper meaning of teaching; cf. Sect. 1.1).

More specifically, diagnosis, properly understood, is knowing about the *probability* (objective) of the presence (or, correspondingly, of the absence) of the illness in question, given the available facts on the case. And by the same token, *correct diagnosis* properly understood is knowing the correct probability of this; it is not a guess about whether a particular illness is present, a guess that may or may not happen to coincide with the truth about this. Only the probability of the presence of the illness is, in principle, generally knowable, as it is only this that the diagnostic profile generally determines (as it only exceptionally is pathognomonic about the presence or absence of the illness; Gr. *patho-*, 'pertaining to illness,' *gnōmonikos*, 'fit to give judgement').

The *correct diagnostic probability* for a given condition is quantitatively coincident with the proportion of instances of the diagnostic profile in general (in the abstract) such that the condition in question is present.

Given that diagnosis is a genre of knowing – a particular type of esoteric ad-hoc (particularistic) knowing – there can be no *'making' of diagnosis*. For different from an artifact, knowing is not made; it is achieved. And diagnostic knowing about a case of sickness is achieved by bringing relevant general knowledge (of medicine) to bear on a relevant set of facts on the case at issue, the appropriate choice of the set of facts also being a matter of general medical knowledge (cf. Sect. 1.1).

Summing up the foregoing, a doctor needs to possess, or else to have ready access to, *general medical knowledge* necessary for the attainment of diagnosis about all of the illnesses explanatory of each of the sicknesses with which patients present for care in his/her discipline of medicine. Regarding any given one of these sickness presentations (s)he is to know: the complete set of elements that constitutes the appropriate clinical profile of the case; the complete set of illnesses and other conditions potentially explanatory of such a presentation (i.e., the differential-diagnostic set corresponding to the type of sickness at issue); the probabilities of each of those illnesses conditional on each of the various possible clinical-diagnostic profiles (commonly enormous in number); the tests that could clinch (practical) rule-in or rule-out diagnosis about each of those conditions; and the diagnostic probabilities corresponding to the various possible post-test profiles (even more numerous than the pre-test ones). This already is, obviously, a very tall order, even though pertaining only to a particular category of the needs for knowledge in diagnostic practice.

Rather than a complaint about a sickness, the prompting of a doctor's diagnostic pursuit can be the client's need for *certification of the absence of some particular illnesses*, for the purposes of employment or insurance, for example. In this situation the doctor need not know the set of illnesses that (s)he is to target for the respective diagnoses about them, ultimately perhaps all of them ruled out; in this situation it is the client who specifies the illnesses that need to be considered.

In respect to the desired rule-out diagnosis about a particular illness, the doctor needs to know the 'risk' indicators and potential manifestations of this illness, and on this basis (s)he needs to document the clinical profile of the case, meaning of the client at the time; and besides, (s)he needs to know what set of laboratory indicators, once their realizations are added to the diagnostic profile, could make the profile pathognomonic about the absence of the illness. (S)he ultimately needs to know the diagnostic probability for the illness in question conditionally on the overall diagnostic profile at the time of the certification.

Different from the client seeking certification of the absence of some particular illness(es) is the one who, while focused on a particular illness, is, for one, expressly free of any possible clinical manifestations (symptoms, overt signs) of that illness; and who is, for another, concerned to possibly receive rule-in diagnosis about that illness, insofar as this illness actually is (latently) present and also detectable (by laboratory-level 'investigation').

This client is seeking the doctor's advice about possibly submitting to *screening* for the illness of his/her concern. As for possible screening for a cancer, the client is one of the 'worried well' who is living in fear of coming down with a clinical case of the cancer and of the clinical course of it that would ensue, possible fatal outcome central among the concerns. And (s)he has hope that the cancer's early, latent-stage detection, followed by correspondingly early treatment, might be curative of the cancer when its late, overt-stage treatment would not be.

Given this client-presentation, the doctor's first-order need is to know the truths that are at issue in the client's fears and hopes: What actually are the probabilities of the client's coming down with a clinical (overt) case of the cancer in various periods of prospective time and age? What actually would be the course of the cancer and what would be the adverse effects of the cancer's treatments if detected in the absence of screening? What actually would be the gain in curability afforded by screening? What would be the adverse effects of the cancer's treatment if diagnosed consequent to the screening, and what would be the harms resulting from the screening?

And then the challenge to the doctor's diagnostic knowledge: what is the probability that a round of the screening, if effected now, would lead to detection (rule-in dgn.) of a case – genuine, life-threatening case – of the cancer?

All of this the doctor (Sect. 1.1) needs to teach the client – for the client to take an informed decision about now submitting to (a round of) the screening.

1.3   The Knowledge Needed for Etiognosis

Once the complaint-prompted diagnostic pursuit has converged to practical ruling-in of a particular illness as the proximal cause of the case of sickness at issue, the doctor may need to get to know about the causal origin – the *etiology/etiogenesis* – of this case of the illness (Gr. *aitia*, 'cause,' 'responsibility'). For, knowing about the causation of the patient's illness – *etiognosis* about the case of illness (see 'gnosis' under diagnosis above) – can be relevant for the management of the diagnosed case of illness.

In this, the first thing the doctor needs to know is the set, the complete set, of known etiologies relevant for the management of the case. Among these (s)he focuses on the ones, if any, for which the patient has a positive history. This is not a differential-etiognostic set, as etiogenetic role for one is not tantamount to absence of it for the others. The concern with any given potential cause in this set is its role in the context of the others in this *etiognostic set* – the possibility that it had a critical role in the genesis of the case of illness (by completing a sufficient cause of it).

Focusing on a given factor in this etiognostic set, the doctor needs to know the set of elements that jointly constitute the *etiognostic profile* of the case, specific to the factor at issue and relevant to the probability that it had a causal role.

Given the etiognostic profile of the case of illness, the doctor needs to know the *probability* that the antecedent in question actually was etiogenetic to the case of the illness at hand. This probability – the *correct* etiognostic probability – is coincident in magnitude with the proportion of instances of the etiognostic profile in general such that the antecedent in it actually is causal to the illness in it.

When the differential-diagnostic set includes potential non-illness causes of the sickness, the need is to achieve etiognosis about these, analogously to etiognosis about illnesses.

The etiognostic demands on a clinician's knowledge are nowhere near as numerous as are the diagnostic ones, but they, when present, generally are distinctly more daunting.

1.4   The Knowledge Needed for Prognosis

Once a particular illness has been identified as the cause of a person's sickness – or when a latent case of a particular illness has been identified as a matter of rule-in diagnosis – and once the etiogenetic work-up, if any, is complete, the doctor's attention turns to the *future course* of this case of illness. (S)he thinks about the possible future manifestations of the illness (in sickness), the possible complications of the illness (its possibly causing other illnesses), and the outcome of the course of the case (fatality, full recovery, recovery with sequelae) insofar as the illness isn't inherently chronic; and these (s)he thinks about conditionally on each of the reasonable options for treatment and perhaps also for future lifestyle, thereby adding potential complications and other adverse effects of treatments to his/her concerns to know about.

So, in respect to those possible future phenomena of health in the context of a rule-in diagnosed case of an illness, the first thing the doctor needs to know is the full set of contextually possible, reasonable *options for treatment*, given the relevant particulars of the case of the illness and also of the patient having this case of the illness.

Regarding the possible future occurrence of any particular one of these phenomena, the doctor needs to know the determinants – the principal ones – of the risk of its future occurrence in the case at hand, conditionally on the treatment at issue. And the ultimate prognostic need is to know the *probabilities* of the phenomena taking place: the probabilities of the event-type phenomena occurring over various spans of prognostic (prospective) time, and of the state-type phenomena being present at various points of prognostic time. Any given one of these probabilities naturally is to be thought of conditionally not only on a particular option for treatment (and possibly also on future lifestyle) but also on the *prognostic profile* of the case at prognostic T_0 (time of prognostication), constituted by the realizations of the accounted-for *prognostic indicators* at that time.

Analogously with correct diagnosis and correct etiognosis, *correct prognosis* about any given event or state is knowing the correct probability of its prospective occurrence (time-specific) conditional on the prognostic profile together with a particular treatment (and possibly lifestyle) prospectively – and also conditional on not succumbing to extraneous causes of death before the time at issue. The correct probability corresponds to the proportion of instances like the one at issue in general such that the event or state (time-specific) will occur.

Like diagnosis and etiognosis, medical prognosis is a genre of knowing, of esoteric particularistic knowing. Thus, different from common parlance in medicine, an illness does not have prognosis, good or bad; the doctor does, the prognosis being good to the extent that it is correct, bad to the extent it is incorrect. And also different from a common notion in medicine, prognosis generally is *not a prediction:* a doctor is predicting a health phenomenon only when expressing a very high probability for this phenomenon's future occurrence.

The concern in prognosis can be about *future occurrence of a new illness*. This can be a topic of prognosis in the context of an existent case of illness, as new illness may arise as a complication of the existent one (e.g., peritonitis from peptic ulcer) or of its treatment (e.g., hemorrhagic stroke from use of an anticoagulant), and the outcome of the course of the existent illness can be a new illness as a sequela (e.g., cirrhosis of the liver following the full course of a case of hepatitis). (Cf. above.)

As a natural extension of this, there is prognosis outside the context of existing illness. It too is a matter of a doctor developing a perception of the probability (correct) that a person free of an overt case of an illness will in the future come down with a case – overt case – of this illness. Thus, when a doctor assesses a woman's risk of coming down with breast cancer (an overt case of it), perhaps in the context of knowing her to be a carrier of a mutation of one of the BRCA genes and on the premise of tamoxifen use to be initiated, (s)he is working toward prognosis about future occurrence of (an overt case of) this illness.

Accordingly, the clinician needs to know the probability of the clinical inception of the illness in question (an overt case of it) for various intervals of prognostic (prospective) time, conditionally on whatever preventive measures might be adopted, and also conditionally on the prognostic profile of the person in relevant respects as well as on not succumbing to extraneous causes of death before the time at issue. These needs are analogous to those in the context of an existent case of an illness, but distinctly simpler in substance and fewer in number.

Chapter 2
The Necessary Forms of the Knowledge

Contents

2.0 Abstract

A patient's presentation for *diagnosis* generally calls for ascertainment (and documentation) of a multidimensional diagnostic profile (Sect. 1.2). In the context of a given type of presentation with sickness (chief complaint, demographic category), the set of diagnostic indicators – a dozen in number, say – definitional to the various possible case profiles generally implies an enormous number of possible profiles for diagnosis about any particular one of the illnesses that could be present, for setting the probability of its presence in the context of the profile at hand. It thus generally is wholly impractical to (learn and) codify the probability for a particular illness separately for each of the possible diagnostic profiles that are possible in the context of a given type of patient presentation.

The need thus is to codify the diagnostic probability for each of the possible illnesses as a *function* – joint function – of the set of diagnostic indicators, defining subdomains of the presentation domain. The idea (in 'clinical epidemiology') that diagnostic probability can be based on the diagnostic 'accuracy' of each of the diagnostic indicators is a serious misunderstanding.

Analogous probability functions are needed for *etiognosis* and *prognosis* just the same.

O. S. Miettinen, *Toward Scientific Medicine*, DOI 10.1007/978-3-319-01671-9_2,
© Springer International Publishing Switzerland 2014

2.1 The Form of Diagnostic Knowledge

The knowledge-base of diagnosis about the presence/absence of possible underlying illnesses in the context of patient presentation with a case of a given type of sickness naturally is to be *organized by types of patient presentation* (chief complaint, demographic category) rather than by types of illness. An example of the categories of the diagnostically relevant knowledge thus is 'the adult presenting with acute dyspnea' and not 'the adult suspected of having pulmonary embolism.'

Given the chief complaint of a person from a given demographic category (broad), the needed knowledge-base is less than daunting in its specification of the set of diagnostic indicators to be addressed for acquisition of the elements in the clinical, pre-lab profile of the case; and the same is also true of the post-lab profile. Nor is the specification of the full differential-diagnostic set dauntingly challenging. But what about the profile-conditional probabilities for the presence of the particular illnesses in the differential-diagnostic set, as the potential causes of this sickness? There generally is an *enormous number of those profiles*, so that knowing the diagnostic probability for any particular illness separately for each of those profiles – those subdomains of the presentation domain – is generally unimaginable; and even if those probabilities were known, profile-by-profile, codification and as-needed accessing of this knowledge would be very impractical.

'Clinical epidemiologists' have been much concerned with the theory of diagnosis for diagnostic practice and diagnostic research. Central in this has been their interest in the 'accuracy' of diagnostic tests with a view to deploying measures of this in the transitions from pre-test probabilities to their corresponding post-test probabilities. And they have extended this thinking by adopting the view that all of the items in a diagnostic profile can be thought of as being results of diagnostic tests, so that the diagnostic probability conditional on the entire profile can be derived by moving successively across all of these 'test' results, starting from a suitable unconditional, 'pre-test' probability. In these terms the number of objects of the needed diagnostic knowledge would be vastly reduced were it not for the fact that this theoretical construct is untenable, flawed in its logic.

This thinking was formerly predicated on the idea that a relevant diagnostic test always produces a *binary result*, either 'positive,' pointing to the presence of the illness in question, or 'negative,' pointing to the absence of the illness. The corresponding relevant properties of a diagnostic test were taken to be the probability that the result will be positive if the illness is present and negative if the illness is absent, the absence meaning that (at least) one of the differential-diagnostic alternatives to the illness in question is present. These two properties got to be termed, respectively, the test's *'sensitivity'* and *'specificity'* for the illness at issue.

A first indication of a concept being malformed commonly is that the corresponding term for it is a misnomer in reference to the concept. In saying that a binary test is 'sensitive' to a particular illness, the meaning ought to be that the result of the test is prone to change from negative to positive on account of the advent of the illness. But in the prevailing meaning of the term, history about myocardial infarction has considerable 'sensitivity' for a fresh case of this illness, without it reacting, in any way, to the new development – without the history in any way 'sensing' the current presence of the illness and turning positive in response to it. The same is true of the patient's age, for example.

And to say that a particular test has good 'specificity' for a fresh myocardial infarction – in that prevailing meaning of a test's 'specificity' to an illness – is downright absurd: practically any diagnostic test, however uninformative about the presence/absence of a fresh MI, is usually negative in its result when there is no fresh MI! In proper terminology one can say, for example, that a positive *result* of the troponin tests is quite specific to the presence of a fresh MI, meaning that its result is positive specifically – solely – in the presence of an active case of myocardial damage (as in a fresh MI), remaining negative in the absence of this; and that a negative result is almost as specific to the absence of this damage.

A test's 'sensitivity' and 'specificity' are also termed its '*true-positive rate*' and '*true-negative rate,*' respectively. But these too are misnomers, as also are their respective antonyms, '*false-positive rate*' and '*false-negative rate.*' In proper terms, a true-positive result of a test is one that truly is positive, regardless of whether the illness in question is present; it is not falsely positive on account of some error in the conduct of the test or in the classification of its result, as would be a false-positive result. And analogously, a true-negative result is, in proper terms, one that truly is negative, again regardless of whether the illness in question is present or absent; it is not falsely negative on account of some error in the conduct of the test or in the classification of its result, as would be a false-negative result.

That these terms are misnomers is particularly problematic for the reason that they are so alluring, so seductive, and therefore so conducive to adoption and perpetuation of the corresponding misunderstandings – of the malformed concepts they represent. So long as critical thought does not interfere, a diagnostician wishes to deploy tests, and so-called tests, that are said to have high 'sensitivity' and high 'specificity' for the illness in question, or high 'true-positive' and 'true-negative' rates – innocently presuming that something meaningful is the denotation of each of these terms.

As the malformed terminology suggests, at issue here indeed are malformed concepts. The *root misconception* here is the notion that a particular test (or other determination) for diagnosis is characterized by particular, singular rates of producing a positive (or negative) result in the presence and absence, respectively, of a particular illness. In truth these rates are, generally, strongly dependent on the pre-test profile. For example, the positivity (and negativity) rates of a troponin test for diagnosis about a fresh MI are, conditionally on the presence of this illness, strongly dependent on the time since the onset of the sickness, which is another necessary item in the diagnostic profile. As another example from the diagnosis about MI, the probability of positive history for MI is highly dependent on the patient's age, in both the presence and the absence of MI, age being another routine element in the diagnostic profile.

When a diagnostic indicator is treated in ordinal or quantitative terms, as indeed is commonly called for, the (malformed) concepts of this indicator's 'sensitivity' and 'specificity' (requiring a cut-off point for the result's classification as positive or negative) are now increasingly replaced, in the teachings of 'clinical epidemiologists,' by *likelihood ratios*: the probability of the test result conditional on the presence of the illness in question divided by its counterpart in the absence of the illness (i.e., in the presence of one of its differential-diagnostic alternatives), separately for each possible result of the test, again for use in transitions from pre-test probabilities to the corresponding post-test probabilities. But just as 'sensitivity' and 'specificity,' these LRs also are not (single-valued) parameters of Nature but just as malformed. Thus, in the context of a binary indicator, the LR for its positive realization is 'sensitivity' divided by the complement of 'specificity,' and for the negative result it is the complement of 'sensitivity' divided by 'specificity.' But given that 'sensitivity' and 'specificity' are not (single-valued) parameters of Nature, this LR isn't either. More on this in Sect. 5.4.

Sadly, this 'sensitivity'-'specificity'-LR conception of the essence of the requisite knowledge-base of diagnosis just continues to live on among 'clinical epidemiologists' and among others affecting understanding of their (malformed) ideas (cf. Sect. 3.1, i.a.).

The fundamental tenable idea about sickness-prompted pursuit of diagnosis is that the *diagnostic profile as a whole serves as a discriminant* between the presence and absence of any particular illness in the differential-diagnostic set. Till the mid-1960s this should have meant a common concern in medical research to develop 'discriminant functions' for diagnosis in such terms as the statistical giant R. A. Fisher had introduced: summary scoring functions based on the diagnostic indicators, so constructed (statistically, from data) that the score's respective distributions in the presence and absence of the illness in question are maximally separated, minimally overlapping.

From the late 1960s on, those concerned to develop the knowledge-base of clinical diagnosis should have been elated to be able to make use of a major innovation in statistics: the introduction of *logistic regression* as a way to produce scoring functions akin the Fisher's discriminant functions but with a major novel property, namely the production of results for the *probabilities* of the two categories in question – the presence and absence of a particular illness, say – conditional on the logistic-type discriminant scores.

As an illustration of potential diagnostic knowledge in the form of a logistic regression function, let us consider again the diagnosis about *myocardial infarction*, MI, in an adult presenting with a complaint about chest pain, CP. And for simplicity, let us take it that the set of diagnostic indicators consists of, only, age, history of MI (positive, negative), type of pain (typical, other), and ST changes in the ECG (present, absent) together with the timing of the ECG as of the onset of the CP.

For the knowledge-base of this diagnosis to take the logistic form, the need is, first, to define a suitable set of *basic statistical variates*. In this case these variates might be taken to be the following:

Y: indicator of MI (i.e., $Y = 1$ if MI present, $Y = 0$ otherwise);

X_1: age in years (i.e., $X_1 = $ age/year);

X_2: indicator of positive history of MI (i.e., $X_2 = 1$ if positive history, $X_2 = 0$ otherwise);

X_3: indicator of CP being typical of MI (i.e., $X_3 = 1$ if typical, $X_3 = 0$ otherwise);

X_4: indicator of ST changes (i.e., $X_4 = 1$ if changes present, $X_4 = 0$ otherwise); and

X_5: time from onset of CP to ECG in hours (i.e., $X_5 = $ time/hour).

Based on these basic variates, a suitable *statistical model* needs to be designed. As one possibility, the adopted statistical model – logistic – could be this:

$$\log[\Pr(Y = 1)/\Pr(Y = 0)] = B_0 + \sum_1^7 B_i X_i,$$

where X_1 through X_5 are as above, while

$X_6 = (1 - X_4)X_5$, and

$X_7 = (X_1)^2$

The role of that X_6 – of the $B_6 X_6$ term in the model – is to provide for the relevance of the absence of ST changes to increase with increasing time from the onset of the CP to the time of the (latest) ECG. (Note that $B_6 < O$.) Addition of X_7 to X_1 provides for the possibility that the probability logit's relation to age is not reasonably linear; that it is better represented by provision for quadratic (parabolic) curvature.

In respect to the CP presentation, a model needs to be designed also for each of the *other illnesses* in the differential diagnostic set. But: the forms of these illness-specific models should differ only in terms of what the dependent variate (Y) indicates (MI, gastroesophageal reflux, ...). The independent variates $(X_1, X_2, ...)$ are to be the same in each of these models, while the values of their associated parameters $(B_0, B_1, ...)$ – these objects of the diagnostic knowledge in respect to the CP presentation – naturally have values specific to the illness the probability of which the function addresses.

This means that *the introductory model above is quite unrealistic.* It is formulated in the spirit of viewing the challenge in the context of a CP presentation to be, simply, discrimination between MI and its absence, when the true challenge is discrimination among all of the illnesses in the differential-diagnostic set. Whereas that model has an indicator for the CP being more-or-less typical of MI (substernal 'aching' or 'pressure'), it should, in the same vein, have an indicator for the CP being more-or-less typical of gastroesophageal reflux (substernal 'burning'), for example.

Perhaps enough has been said here to illustrate the point that for sickness-prompted diagnosis the appropriate and *necessary form of the knowledge-base* is that of *diagnostic probability functions* separately for each of the various types of patient presentation, each of the functions specifying the various profile-conditional probabilities for a particular illness being present (and explanatory of the sickness), accomplishing this for a myriad of possible profiles in terms of the known values of quite a modest number of parameters in a suitably designed statistical model.

Given the criticism I presented about deriving the diagnostic probabilities by sequential application of indicator-specific pseudo-constant likelihood ratios to the realizations of the indicators – which ignores the correlatedness of the indicators (and leaves the unconditional 'pre-test' probability unspecified) – it is important to appreciate that the parameter (B_i) for any given independent variate (X_i) in a logistic (or other) probability function is *inherently conditional* on all of the other independent variates in it. The parameter associated with a given independent variate would have a different value if the set of other independent variates in the function were different; and whatever the value of a given one of these parameters is in the context of whatever set of independent variates in the functions, its value fully reflects the correlations – relative redundancies – among those variates.

Much less complex than the knowledge-base of sickness-prompted pursuit of diagnosis is that of mere pursuit of *rule-out* diagnosis of a particular illness specified a priori (Sect. 1.2). For any given one of the illnesses whose absence the client wishes to get certified, the doctor needs to have access to perhaps only one function – logistic, with some laboratory determinant(s) eminent – such that the diagnostic probability implied by it can be sufficiently conclusive about the absence of the illness in question. If the probability from this function is not low enough, (s)he needs to know what other test result(s) added to the profile has the potential to yield a probability that is sufficiently low; and (s)he needs to have access to that added post-test function.

For the purposes of decisions about *screening* for a cancer (Sect. 1.2) the relevant diagnostic knowledge has to do with a particular algorithm for pursuing *rule-in* diagnosis about the cancer – the algorithm of choice in the place and at the time in question. Regarding the client at hand the diagnostic need is to know the probability that this regimen's application, on this person at this time, would provide for rule-in diagnosis about the cancer – specifically Stage I diagnosis about a genuine case of it (Sect. 1.2).

The needed diagnostic knowledge is of the form of a probability function for the attainment of the sought-for Stage I rule-in diagnosis by the application of the regimen of choice. It is a function based on a logistic model, with its independent variates based on indicators of 'risk for' (really of the probability of latent presence of) the cancer. If at issue is possible repeat screening – with a regimen specific to this – an added determinant of the probability is time since the previous round of the screening (with negative result of the diagnostic pursuit).

2.2 The Form of Etiognostic Knowledge

When the differential-diagnostic set includes a non-illness (a medication use, say) as a potential cause of the patient's sickness, the doctor may turn to etiognosis about this potential cause of the sickness.

And when the identified cause of the sickness is an illness and the proper management of this illness is dependent on whether a particular antecedent of the case that was there actually was causal to the case, (s)he is concerned to know the etiognostic probability for that antecedent, for its etiogenetic role.

For etiognosis it isn't really necessary to have (access to) an actual etiognostic probability function, one which for a domain of the case's occurrence expresses the etiognostic probability for the antecedent as a function of particulars of that history and perhaps of the case of illness also, together with whatever else goes into the etiognostic profile (Sect. 1.3).

The reason for the essential redundancy of that EPF is that each of those probabilities is determined by a related measure, the corresponding *etiogenetic rate-ratio*: the etiogenetic probability for a particular antecedent that was there is

$(RR - 1)/RR,$

where *RR* is the etiogenetic rate-ratio specific to the etiognostic profile of the case. Thus the requisite knowledge can be of the form of suitable *RR functions*, each specific to a particular type of illness, or sickness in the absence of illness, together with a particular type of potentially etiognenetic antecedent of it.

2.3   The Form of Prognostic Knowledge

What I said about the requisite knowledge-base for prognosis about the illness that has been identified as the cause of a case of sickness (Sect. 1.4) and about the necessary form of the knowledge-base for diagnosis about sickness-explanatory illness (Sect. 2.1) implies by analogy the necessary general form of the knowledge-base for prognosis in the context of an existent, rule-in diagnosed case of an illness.

While the knowledge-base for sickness-prompted pursuits of diagnoses needs to be organized by types of sickness in the patient presentations (Sect. 2.1), that for prognoses in the contexts of existent, rule-in diagnosed cases of illness, like that for etiognoses, needs to be organized by types of illness. In the context of any given type of illness the relevant prognoses are about possible prospective overt manifestations and complications of the illness and about its possible outcomes, and also about possible adverse effects of treatments of the illness (Sect. 1.4), but commonly not about the illness (somatic anomaly) per se.

In respect to the probability of a given one of the possible prospective phenomena actually occurring, the counterpart of a diagnostic probability function is a *prognostic probability function*. For a given type of prospective *event*, such as death from the illness, the PPF expresses the probability (cumulative) of its occurrence as a function of prospective/prognostic time jointly with prognostic indicators at prognostic T_0 (descriptive of the case of illness as well as of its 'host,' the patient); and the determinants in the function generally also include prospective treatment and potential modifiers of its effect – all of these again expressed in terms of statistical variates (Xs, numerical). If at issue is a *state* of health, such as chronic coronary heart disease (coronary stenosis manifest in propensity for episodic, exertion-prompted angina pectoris), a corresponding function addresses the probability of its being present at a given point in prognostic time, in otherwise similar terms. These PPFs would generally be conditional also on not having succumbed to an extraneous cause of death.

The needed PPFs are generally logistic for state-type phenomena, but for event-type phenomena they are logistic only for a very short time horizon (within which the particulars of timing are immaterial and deaths from extraneous causes are too rare to consider) (Sect. 7.3).

The requisite knowledge-base for prognosis about *future case of an illness* – an overt case of it – in a person free of it is quite analogous to that of prognosis about future phenomena in an existent case of an illness, only distinctly less complex (Sect. 1.4). And by extension, the necessary form of this knowledge across a multitude of types of risk profile also is analogous to the counterparts of this in the context of an existent case of an illness – only, again, distinctly less complex (Sect. 1.4).

Chapter 3
The Knowledge According to Its Source

Contents

3.0 Abstract

In contemporary *textbooks* of "medicine" there is a complete absence of content on general concepts and principles – general theory – of medicine, or there is merely a token treatment of topics that are misconstrued as being ones of general principles of medicine. The substantive contents of these general textbooks, and also those of particular disciplines of medicine, are not organized with a view to the knowledge needs for diagnosis and prognosis, respectively; and etiognosis never is even a topic. Very little of the contents addresses diagnostic or prognostic probabilities, and this, even, generally amounts only to a few very sketchy remarks. And even the very concepts of common illnesses are liable to remain remarkably muddled.

Medical *journals*, too, are poor sources of genuine knowledge relevant for the practice of medicine, while rich, instead, in reports on and unjustifiable conclusions and pseudo-conclusions from poorly conceived and poorly reported studies.

In this context, unsurprisingly, medical *education* remains largely devoid of well-conceived purposes and relevant contents.

So, the entire knowledge culture of medicine is in need of radical overhaul and rectification.

O. S. Miettinen, *Toward Scientific Medicine*, DOI 10.1007/978-3-319-01671-9_3,
© Springer International Publishing Switzerland 2014

3.1 General Textbooks on the Needed Knowledge

Chapter 1 here was dense with *general principles of medicine*, as it laid out what a doctor in principle (*sic*) needs to know, as a matter of general medical knowledge, when caring for a client in a given one of several generic types of encounter with them. And implicit was the larger principle that a doctor should not only know what it is that (s)he in principle needs to know in a client encounter of a given generic type but that (s)he then needs to actually bring the requisite general medical knowledge to bear on the acquired relevant particularistic facts on the case – to thereby achieve professional, esoteric knowing (probabilistic) about the client's health: diagnosis, etiognosis, and/or prognosis. Beyond these, there were the principles that the doctor should, whenever possible, teach the client about their health (about the attained gnosis about it), and that decisions about actions (testings, interventions, lifestyles) should be left to the thus-informed client whenever possible (and desired by the client).

Regarding the *general medical knowledge*, the deployment of which the principles call for, it was evident from Chap. 1 that its natural organization for diagnosis is by type of client presentation, and that organization by type of illness is natural only for the purposes of etiognosis and prognosis. And it also was evident from Chap. 1 that the requisite general medical knowledge is very complex, so that, as set forth in Chap. 2, the only practical form for codification of the requisite knowledge-base of medicine is that of *gnostic probability functions*.

The contemporary knowledge-base of medicine (gnosis in it) is codified in current textbooks of medicine. Some of them need to be general in the sense of dealing with medicine at large and, consequently, rather superficially – for the purposes of primary care, provided by that discipline of medicine which has the principal role in the triage of clients to other disciplines of medicine (should such referral be needed). Various other textbooks need to be directed to the many medical disciplines that are narrower and deeper in their concerns. (It is rather unnatural to speak of 'general practice' and 'specialties' of modern medicine, given that a modern generalist of medicine is unimaginable – just as there are seen to be no generalists, and hence no specialists, in the professions of engineering, music, and sports, for example.).

For illustration of the present state of the codification of the knowledge-base of general, primary-care medicine, I here consider three eminent textbooks concerned with it, one from the time before medicine's differentiation into its modern subdisciplines (ref. 1 below) and two strictly contemporary (refs. 2, 3):

References:

1. Osler W (1892) The principles and practice of medicine: designed for the use of practitioners and students of medicine. D. Appleton and Company, New York

2. Colledge NR, Walker BR, Ralston SH (eds) (2010) Davidson's principles and practice of medicine, 21st edn. Churchill Livingstone, Edinburgh
3. Goldman L, Ausiello D (eds) (2008) Cecil Medicine, 23rd edn. Saunders/ Elsevier, Philadelphia

Osler's book – a true classic in medicine – has no Introduction or Preface, nor does it elsewhere explicate the duality in its title constituted by the principles of medicine on one hand and the practice of medicine on the other hand; but implied by its title is that the book is about both of these two aspects of medicine, and that it is to be used both as a textbook in the study of medicine and as a handbook in the practice of medicine, the unitary discipline of medicine. Over 1,000 pages in length, this single-author book is divided into 11 'sections,' 10 of them according to type of illness. The respective section titles are: (1) Specific infectious diseases; (2) Constitutional diseases; (3) Disease of the digestive system; (4) Diseases of the respiratory system; (5) Diseases of the circulatory system; (6) Diseases of the blood and ductless glands; (7) Diseases of the kidneys; (8) Diseases of the nervous system; (9) Diseases of the muscles; (10) The intoxications, sun-stroke; obesity; and (11) Diseases due to the animal parasites.

As an illustration of the specific content in this book, let us examine under "Diseases of the circulatory system" the subsection entitled "Neuroses of the heart" and in it the content concerning the "neurotic disease" *Angina pectoris,* there being no section on myocardial infarction (under whatever term for this). The opening paragraph is this:

Stenocardia . . . is not an independent affection, but a symptom associated with a number of morbid conditions of the heart vessels, more particularly with sclerosis of the root of the aorta and changes in the coronary arteries. True angina, which is a rare *disease*, is characterized by paroxysms of agonizing pain in the region of the heart, extending into the arms and neck. In violent attacks there is a sensation of impending death. [Italics added.]

So, angina pectoris – different from stenocardia – is introduced as a "disease." And the text that follows indeed addresses angina pectoris not as a symptom but as a disease, as the topics are "Etiology and pathology," etc., as is routine in characterizations of diseases.

Under "Symptoms" of this "disease" are descriptions of the anginal pain and its prompting, and of its associated other symptoms and also signs; and under "Diagnosis" the content begins with this:

In true angina, even in the milder forms, signs of arterio-sclerosis are usually present. In a case presenting attacks of praecordial pain or pains in the cervical or brachial plexuses, if the aortic second sound is clear, not ringing, the pulse tension low, and the peripheral arteries soft, the diagnosis of true angina should not be made. After all, the chief difficulty, however, arises in the cases of the hysterical or *pseudo-angina*. [Italics in the original.]

A paragraph on pseudo-angina follows, and the third and final paragraph under "Diagnosis" is this:

> There are cases, in women, which are sometimes very puzzling; for instance, when the patient presents a combination of marked hysterical manifestations and attacks of angina and has aortic insufficiency. In such instances the patient should receive the benefit of the doubt and be treated for true angina.

Under "Prognosis" the text is, in its entirety, this:

> Cardiac pain without evidence of arterio-sclerosis or valve disease is not of much moment. True angina is almost invariably associated with marked cardiovascular lesions in which the prognosis is always grave. With judicious treatment the attacks, however, may be long deferred, and a few instances recover completely. The prognosis is naturally more serious with aortic insufficiency and advanced arterio-sclerosis. Patients who had well-marked attacks may live for many years, but much depends upon the care with which they regulate their daily life.

Thus, Osler evidently did not have the disease concepts of coronary stenosis and coronary thrombosis, much less of them as causes of myocardial ischemia, and of angina pectoris as a clinical manifestation – symptom – of the latter. For this reason, angina pectoris seemingly was, to him, a clinical syndrome signifying an unknown somatic anomaly; and its diagnosis was a matter of the syndrome's 'pattern recognition' in conjunction with some other indications of that anomaly's underlying arterio-sclerotic disease (not otherwise specified).

Even though it is not quite clear from this example, diagnosis to Osler appears to have been the illness the patient gets to be presumed to have, something that is "made" into this presumption; it was not the probability with which a particular type of somatic anomaly is seen to be present (cf. Sect. 1.2).

And prognosis to Osler apparently was the valuation of the presumptive prospective course of the patient's illness – characterized as "grave," "serious," etc. – rather than the doctor's knowing about the prospective course as a matter of probability for a given variant of it (which knowing cannot be characterized as, e.g., "grave").

Written before medicine got to be differentiated into its component disciplines, Osler's book naturally was directed to all physicians – as was made explicit in its subtitle ("Designed for the use of practitioners and students of medicine," without exceptions). It thus was a textbook of all of medicine.

It deserves particular note that Osler's classic textbook of medicine was organized by types of illness ("disease") within the organ system affected and not at all by types of patient (or other client) presentation. (Cf. Sects. 2.1, 2.2 and 2.3.)

The *Davidson's* textbook (ref. 2 above) is akin to Osler's in that it too has the "principles and practice" duality in the title of it; and it too has no Preface. But it differs from Osler's book by not stating who the intended readers are, and by actually having that principles-practice duality clearly manifest in the organization of its contents.

In that book, despite its title and also content under the rubric of "*Principles of medicine,*" there is no conception of principles of medicine of the types that were addressed in Chaps. 1 and 2 here, and were touched upon also in the beginning of this chapter. Part 1 in that book, with this title, consists of seven chapters, with these the respective headings: (1) Good medical practices; (2) Therapeutics and good prescribing; (3) Molecular and genetic factors in disease; (4) Immunological factors in disease; (5) Environmental and nutritional factors in disease; (6) Principles of infectious disease; and (7) Ageing and disease.

That first chapter – on "Good medical practices" – consists of merely uncritical description of, for example, the concepts of a diagnostic test's 'sensitivity' and 'specificity,' which I characterize as serious misapprehensions in Sects. 2.1 and 5.4 of this book. As another example, it describes, uncritically, the teachings concerning Evidence-based Medicine, which I have extensively criticized in another book (ref. 3 in Preface):

To me, *good medical practices* are ones in conformity with the imperatives of medical professionalism. They are founded on and constrained by logically and ethically tenable principles of medicine. They are maximal possible deployment of genuine knowledge (substantive) in this framework (to set gnostic probabilities), and on this basis teaching the client about their own health. To me, good medical practices are what principles of medicine are all about, first generically and then in reference to particular contexts. (Cf. Chaps. 1 and 2 above.) And to me, thus, none of those seven chapters under "Principles of medicine" actually is about principles of medicine.

Part 2 of the book – on "*Practice of medicine*" – conforms to the structure of the entire book of Osler's: its chapters address particular types of illness entity and say next to nothing about practice, much less about good medical practices, specific to each of these.

As for *angina pectoris*, in a chapter by D. E. Newby, N. R. Grubb, and A. Bradbury on "Cardiovascular disease" the authors state that "Angina pectoris is the symptom complex caused by transient myocardial ischemia and constitutes a clinical syndrome rather than a disease." And related to this, they say that "*Acute coronary syndrome* is a term that encompasses both unstable angina and MI [myocardial infarction]," and they address this "syndrome" – actually the union of a "symptom complex" and a disease – rather like the way Osler addressed angina pectoris alone when describing it as a disease: describing its "Clinical features," "Diagnosis and risk stratification," "Investigations," and "Immediate management: the first 12 hours" followed by the later counterpart of this.

Under "Clinical features" of this "ACS" the authors characterize "the cardinal symptom" of it (of this union of a symptom and a disease) – "Prolonged cardiac pain: chest, throat, arms, epigastrium or back" – but then point out the existence of variable features, which actually do not belong in the "syndrome" as they define it (above):

> Most patients are breathless and in some this is the only symptom. Indeed, MI may pass unrecognized. Painless or 'silent' MI is particularly common in older patients or those with diabetes mellitus. If syncope occurs, it is usually due to an arrhythmia or profound hypotension. Vomiting and sinus bradycardia are often due to vagal stimulation and are particularly common in patients with inferior MI. Nausea and vomiting may also be caused or aggravated by opiates given for pain relief. Sometimes infarction recurs in the absence of physical signs.

The remaining, third paragraph on the clinical features of this "syndrome" is, quite incongruously, about sudden death and cardiac failure.

Regarding diagnosis about this "syndrome," said is only this:

> The differential diagnosis is wide and includes most causes of central chest pain or collapse (pp. 535 and 552). The assessment of acute chest pain depends heavily on an analysis of the character of the pain and its associated features, evaluation of the ECG, and serial measurements of medical markers of cardiac damage, such as troponin I and T.

A section on "Investigations" follows, addressing the ECG and other tests, with no reference to their 'accuracy' in terms of 'sensitivity,' etc. (addressed under "Principles of medicine"). No diagnostic probabilities are given. But: at issue really is diagnosis about acute coronary heart disease (ischemic, possibly involving myocardial damage) rather than "diagnosis" of a syndrome (mere pattern recognition of manifestations).

Associated with the paragraph on diagnosis is another on "Risk stratification," meaning prognosis, and not of any syndrome but of acute coronary heart disease:

> Approximately 12% of patients will die within 1 month and a fifth within 6 months of the index event. The risk markers that are indicative of an adverse prognosis include recurrent ischemia, extensive ECG changes at rest and during pain, the release of biochemical markers (creatinine kinase or troponin), arrhythmias, recurrent ischemia and haemodynamic complications (e.g., ...) during episodes of ischemia.

I am at a loss to understand why the "ACS" term has been adopted for what now (different from Osler's time) is understood not to be a syndrome – a complex of clinical manifestations (symptoms and/or signs) – but an actual illness: *acute myocardial ischemia* secondary to a fresh coronary thrombosis that constitutes an acute, partial or total, coronary occlusion (a total occlusion leading from myocardial ischemia to myocardial infarction and, in nonfatal cases of this, ultimately to myocardial fibrosis). The "ACS" term is a misnomer for what is now understood to be *acute coronary heart disease* in this meaning. (The fresh, partially occluding thrombosis is not angina, nor is it inherently unstable.)

For two other examples of how the "practice" – really knowledge-base (cf. above) – of medicine is addressed in the current edition of *Davidson's*, I turn to the chapter on "Respiratory disease," by P. T. Reid and J. A. Innes.

First *pneumonia*. The text begins with this:

> Pneumonia is defined as an acute respiratory illness associated with recently developed radiological pulmonary shadowing which may be segmental, lobar or multilobar.

It thus is presented as an acute respiratory illness with no defining pathological particulars to distinguish it from pulmonary embolism, for example; and the "definition" involves a (vague) diagnostic criterion!

Under community-acquired pneumonia (distinct from hospital-acquired), "Clinical features" are described as to their variability but without any relative frequencies for the described variants, and with no probabilities of the presence of pneumonia conditional on these.

A section on "Investigations" follows, opening with this: "The objectives are to exclude other conditions that mimic pneumonia (Box 19.44) [the title of this Box, listing "pulmonary infarction" among others, is "Differential diagnosis of pneumonia"], assess severity, and identify the development of complications." But nothing is said about the way in which those possible other conditions that mimic pneumonia would be excluded. Said is only – without specifics – that "A chest X-ray usually provides confirmation of the diagnosis." There is, thus, *no statement of any diagnostic probability*. Nor is there any specification of a test's 'sensitivity,' etc., despite this being a topic in the book's very first chapter, on "Good medical practices." (Cf. angina pectoris above.)

There is a single paragraph on "Prognosis," opening this way: "Most patients respond promptly to antibiotic therapy. However, fever may persist for several days and the chest X-ray often takes several weeks or even months to resolve, especially in old age." *No prognostic probabilities are given.*

As the other example of respiratory disease I take up *pulmonary embolism*, which in this book is addressed in a chapter on "Venous thromboembolism" (VTE), by P. T. Reid and J. A. Innes. Under "Clinical features" is said only this:

> VTE is difficult to diagnose. It may be helpful to consider the following:
>
> – Is the clinical presentation consistent with PE?
> – Does the patient have risk factors for PE?
> – Are there any alternative diagnoses that can explain the patient's presentation?
>
>   Presentation varies depending on the number, size and distribution of emboli, and the underlying cardiorespiratory reserve (Box 19.95). [That Box addresses "Features of pulmonary thromboemboli" in respect to pathology, symptoms, signs, chest X-ray, ECG, arterial blood gases, and alternative diagnoses, separately for acute massive PE, acute small/medium PE, and chronic PE.] A recognized risk factor is present in between 80 % and 90 % of patients (Box 19.96). The presence of one or more risk factors may multiply the risk.

Brief statements of various "Investigations" follow, with no word about the 'accuracy' of any of these. There is *no section, nor otherwise any word, on diagnosis*.

A section on "Management" is followed by two brief paragraphs on "Prognosis":

Patients who have suffered symptomatic VTE carry an increased risk of further events, particularly if a persisting risk factor is present. The risk of recurrence is highest in the first 6–12 months after the initial event, and by follow-up at 10 years just under one-third of patients may have suffered a further event. Risks of recurrent VTE are lowest in those with temporary or reversible risk factors.

The immediate mortality is highest in those with echocardiographic evidence of right ventricular dysfunction or cardiogenic shock. The vast majority of patients attain normal right heart function by 3 weeks but persisting pulmonary hypertension may be present in around 4% of patients by 2 years. A minority progress to overt right ventricular failure.

Davidson's thus appears to be quite an inadequate source not only of "Principles of medicine" but also of practice-relevant (gnosis-serving) knowledge about acute coronary heart disease, pneumonia, and pulmonary embolism as topics for "Practice of Medicine" – whether general primary-care medicine or the differentiated medicine of cardiologists and pulmonologists. Like the classic text of Osler's (ref. 1 above), this book is organized by types of illness and not at all by types of client presentation for diagnosis (Sect. 1.2).

The *Cecil* textbook (ref. 3 above) is substantially more extensive than *Davidson's*, consisting of almost 3,000 pages. It does have a Preface, and in it the Editors say this as the essential introduction to the book:

The 23rd Edition of the *Cecil Medicine* ... [represents] the latest medical knowledge ...

The contents of *Cecil* have remained true to the tradition of comprehensive textbook of medicine that carefully explains the *why* (the underlying normal physiology and pathophysiology of disease, now at the cellular and molecular as well as the organ level) and the *how* (now frequently based on Grade A evidence from randomized controlled trials). ...

The sections for each organ system begin with a chapter that summarizes an approach to patients with key symptoms, signs, or laboratory abnormalities associated with dysfunction of that organ system. As summarized in Table 1-1, the text specifically provides clear, concise information on how a physician should approach over 100 common symptoms, signs, and laboratory abnormalities, usually with a flow diagram and/or table for easy reference. ...

The Preface does not say who, at present still, are the intended readers of a book that has "remained true to the [Oslerian] tradition of comprehensive textbook of medicine."

This 3,000-page book – representing work by five Associate Editors, two Coordinating Editors, some 400 Contributors, and a Global Advisory Board of 30 members – is divided into 28 Parts: (1) Social and ethical issues in medicine; (2) Principles of evaluation and management; (3) Preventive and environmental medicine; (4) Aging and geriatric medicine; (5) Clinical pharmacology;

(6) Genetics; (7) Principles of immunology and inflammation; (8) Cardiovascular disease; (9) Respiratory disease; ...; (27) Skin diseases; (28) Reference intervals and laboratory values.

Before focusing again on angina pectoris, pneumonia, and pulmonary embolism, some notes on the big picture may be in order here. In Chap. 1 (of Part 1) the editors write about "Medicine as a learned and humane profession." This Chapter opens with "Approach to medicine," in which the first sentence is this: "Medicine is a profession that incorporates science and the scientific method with the art of being a physician"; but the last paragraph opens with the incontrovertible negation of this unitary conception of medicine: "The explosion of medical knowledge has led to increasing specialization and subspecialization, ...," meaning that medicine actually is the aggregate of an increasing multitude of particular disciplines of medicine; and so the question really is: Who are the intended readers of this "comprehensive textbook of medicine"? Also, those disciplines (differentiated arts) of medicine do not incorporate science; they only increasingly apply scientific knowledge. And the very existence of "the scientific method" has remained a matter of controversy within actual science (ref. below), and I am unaware of any serious, learned attempt at the definition of this for the practice of (the arts of) medicine.

Reference: Blake RM, Ducasse CJ, Madden EH (2003) Method, scientific. In: Heilbron JL (ed) The Oxford companion to the history of modern science. Oxford University Press, Oxford, pp 519–520

Immediately preceding the Chap. 70 on *angina pectoris* is one on "Atherosclerosis, thrombosis, and vascular biology," by V. Fuster. Under its final subheading "Coronary atherothrombotic disease" said is that this "includes a wide spectrum of conditions, ranging from silent ischemia and exertion-induced angina (Chap. 70) to the acute coronary syndromes (Chap. 71)." That Chap. 71 is, however, entitled "Acute coronary syndrome: ..." (singular).

The Chap. 70 entitled "Angina pectoris," by P. Théroux, opens with the subheading of "Definition" and in it with the statement that "Angina is the most frequent clinical expression of myocardial ischemia"; but it never makes it clear that the focus is on exertion-induced angina (cf. above). The ensuing subsection on "Epidemiology" (under "Angina pectoris") actually is about the epidemiology of "coronary artery disease" and "coronary heart disease," both of them left undefined. The last sentence in this subsection is about "the risk of coronary artery disease, MI, and angina," suggesting that angina pectoris actually is a disease unto itself (rather than only a "clinical expression of myocardial ischemia"). But the subsection on "Clinical manifestations" suggests that the Chapter really is about myocardial ischemia, as this subsection is about the manifestations of such ischemia, in various patterns of angina. This subsection introduces the distinction between "stable" and "unstable" angina, but nowhere is it said that this Chapter is specifically about stable angina, even though this focus becomes evident later in the Chapter.

In the subsection on "Diagnosis" the concern is with "evaluation" of a patient with "chest pain," with a view to "the likelihood of CHD [coronary heart disease]." A table addresses "Approximate sensitivity and specificity of common tests to diagnose coronary artery disease" – as though these measures of the tests' 'accuracy' had values independent of the patient's pretest profile (see Sects. 2.1 and 5.4 of this book). A Figure depicts "Approximate probability of coronary artery [*sic*] disease before and after noninvasive testing in a patient with typical angina pectoris," but the way of setting the probability of CAD (undefined) on the basis of the wide variety of profiles before and after those testings is left unspecified.

A subsection on "Treatment" follows, opening with the point that "The management of angina pectoris aims to prevent death and MI, reducing ischemic episodes to improve quality of life and slowing or even reversing the process of atherosclerosis." But at issue actually is not management of the angina pectoris symptom (of myocardial ischemia, palliatively) but treatment of its underlying *chronic CHD* – coronary stenosis manifest in exertion-induced angina pectoris.

As for prognosis, a table addresses "[CAD] prognostic index" in terms of specifying 5-year mortality rate under medical treatment – undefined – alone, by "Extent of [CAD]," from "1-vessel disease, 75 %" to "3-vessel disease, ≥ 95 % proximal [left anterior descending coronary artery]."

This confusing Chapter, I suggest, would have been much cleaner and clearer had the topic been presented not as angina pectoris but as *chronic CHD*, CCHD.

In like manner and consistent with this, the ensuing Chap. 71, by D. D. Waters, would have been much clearer, I suggest, had it been said to be about *acute CHD*, ACHD, rather than about "*Acute coronary syndrome: unstable angina and non-ST segment elevation myocardial infarction.*" ACHD would have been understood to be acute (thrombotic) increase in coronary stenosis, manifest (acutely) in new, or lower threshold for, exertion-associated angina pectoris or even in angina pectoris at rest – corresponding to partial and complete occlusion of a coronary artery, respectively, or to impending and actual MI, respectively. These clear concepts and terms used to be well-established, with no concept of syndrome involved.

Under "Definition" is first said that "Unstable angina is distinguished from stable angina (Chapter 70) by the new onset or worsening symptoms [*sic*] in the previous 60 days [*sic*] or by development of post-MI angina 24 hours or more after the onset of MI." But to me the latter is a special case of the former. The text continues: "When the clinical picture of unstable angina is accompanied by elevated markers of myocardial injury, such as troponins or cardiac isoenzymes, non-ST segment elevation MI is diagnosed." So, this subtype of MI appears to be diagnosed without regard for the presence/absence of ST elevation, even though the latter is in the very term for this subtype of MI. That awkward term, I take it, actually refers to relatively minor, *nontransmural* MI, distinct from "massive," transmural MI.

The *syndrome* at issue here actually is the triad (manifestational) consisting of acutely new or acutely worsened angina together with elevated cardiac enzymes but without ST elevation in the ECG; and this syndrome is understood to be pathognomonic about the presence of its underlying *illness* (somatic anomaly), about non-transmural (rather than transmural) MI. This, I suggest, is a genuine ACS of the non-ST-elevation subtype, to be distinguished from ACS involving ST elevation (and signifying transmural MI).

Under "Diagnosis" – purportedly about that so-called ACS but actually about myocardial ischemia and infarction – said is that "The initial assessment should be directed toward determination of whether the symptoms are caused by myocardial ischemia and, if so, the level of risk [*sic*]. The probability of MI can be estimated from the history, physical examination, and electrocardiography (Fig. 71-1). This information and the assessment of the patient's clinical features should indicate whether the probability that the symptoms are due to myocardial ischemia is high, intermediate, or low (Table 71-2)." Examples from that table, addressing "Likelihood that unstable angina symptoms are caused by myocardial ischemia," are these: "High likelihood given any of the following features: . . ., definite angina in men $\geq$ 60 years or women $\geq$ 70 years, . . ." But, in point of a-priori fact, angina implies presence of myocardial ischemia unconditionally, by its very definition, and at issue was supposed to be the likelihood (probability) of MI rather than mere ischemia of the myocardium.

That Fig. 71-1 is entitled "Flow diagram to estimate the risk [*sic*] of acute myocardial infarction (MI) in emergency departments [*sic*] in patients with acute chest pain," falsely implying that the same diagnostic profiles in other settings would imply different probabilities. Translated into a logistic function, that "flow diagram" defines the log-odds of MI for a patient presenting with chest pain as:

$$\text{Log } [P/(1 - P)] = 1.15\,X_1 - 0.94(1 - X_1)X_2 - 3.89(1 - X_1)(1 - X_2)X_3$$
$$- 2.09(1 - X_1)(1 - X_2)(1 - X_3)X_4X_5 \ldots$$

where

P = probability of MI;

X_1 = indicator of "ST elevation ($\geq$1 mm or Q waves ($\geq$0.04 sec) [*sic*] in 2 or more leads, not known to be old" (i.e., $X_1 = 1$ if this, $X_1 = 0$ otherwise);

X_2 = indicator of "ST-T changes of ischemia or stain [*sic*], not known to be old";

X_3 = indicator of "Time since chest pain began $\geq$ 48 hour";

X_4 = indicator of "Prior history of angina or MI";

X_5 = indicator of "Pain worse than usual angina or the same as prior MI";

etc.

Thus, with $X_1 = 1$, $P = \exp (1.15)/[1 + \exp 1.15] = 0.76$ (as in Fig. 71-1); with $X_1 = 0$ and $X_2 = 1$, $P = \exp (-0.94)/[1 + \exp (-0.94)] = 0.28$; with $X_1 = 0$,

$X_2 = 0$, and $X_3 = 1$ the value of P is 0.02; with $X_1 = X_2 = X_3 = 0$ and $X_4 = X_5 = 1$, P is 0.11 – all of these as in Fig. 71.1.

This formulation of the probability of MI in the context of "acute chest pain" (implied by Fig. 71-1) is very *ill-conceived*. Suffice it to point out the absence of any further characterization of the "acute chest pain" and the meaninglessness of that X_5 in the absence of prior CHD.

Addressed under "Prognosis" is not knowledge about the prospective course of cases of "ACS." Some statistics in this section focus on "unstable angina," concerning its associated occurrence of MI and of death, but with no distinction-making according to prognostic indicators.

There is a separate Chap. 72 on "ST segment elevation acute myocardial infarction and complications of myocardial infarction." However, both pathologically defined types of MI, nontransmural and transmural, could – and should – have been addressed under acute CHD, along with impending/imminent MI, coherently and much less confusingly.

Pneumonia is addressed in a Chap. (97) entitled "Overview of pneumonia," by A. H. Limper. The definition of pneumonia in this *Cecil*, very different from *Davidson's*, is a matter of type of pathology free of any reference to radiography. Our focus here, as before, is on community-acquired pneumonia, and it is defined as "infectious pneumonia in patients living independently in the community."

Under "Clinical manifestations" the orientational point is that "The possibility of pneumonia should be considered in any patient who has new respiratory symptoms, including cough, sputum, or dyspnea, particularly when these symptoms are accompanied by fever or abnormalities on physical examination of the chest, such as rhonchi and rales." And the main points are these: "The classic physical findings in lobar pneumonia include evidence of consolidation with altered transmission of breath sounds, egophony, crackles, and changes in tactile fremitus. However, in many patients, the physical findings are more subtle and may be limited to scattered rhonchi. A thorough physical examination, posterioranterior and lateral chest radiographs, and blood leucocyte count with differential call count *should* be performed when pneumonia is suspected. An assessment of gas exchange *should* be obtained for all patients who are admitted to the hospital. The clinician needs to be mindful of competing diagnoses [*sic*] that can mimic the findings of pneumonia, such as pulmonary embolism (Chapter 99), . . ." (Italics added.)

After all of this complexity and vagueness about the manifestations of community-acquired pneumonia comes a subsection on "Diagnosis." In it, remarkably, there is nothing at all about the diagnostic probability (of the presence of pneumonia), given the patient's diagnostic profile, nor about the 'accuracies' of the various diagnostic tests (addressing potential non-clinical manifestations of the illness) that "should" be carried out. Subsections on "Prevention" and "Treatment" follow; but there is no subsection on prognosis, nor is this addressed under Treatment.

The Chap. (99) on *Pulmonary embolism*, by V. F. Tapson, naturally is mainly about thrombotic PE. Definitionally, in line with *Davidson's* (above), it makes the point that "Thrombus from the deep veins of the lower extremities is by far the most common material to embolize the lungs; deep vein thrombosis (DVT) and PE must be recognized as parts of the continuum of one disease entity, venous thromboembolism (VTE)." But: a Table (99-1) on "Symptoms and signs in patients with acute pulmonary embolism [there is no chronic PE] without preexisting cardiac or pulmonary disease" declares that only 11% of PE patients have DVT.

"The history and physical examination are notoriously insensitive and nonspecific for both DVT and PE." (Cf. 'sensitivity' and 'specificity' in Sect. 2.1.) Under "Diagnosis" is first presented a Table entitled "Differential diagnoses of acute pulmonary embolism," described in the text as specifying "Diagnoses [*sic*] that are commonly present with chest pain or dyspnea and, in a few cases, hemoptysis, and that might be considered along with acute [*sic*] pulmonary embolism, depending on the clinical setting [*sic*]." (They include: myocardial infarction, pericarditis, tachycardia (>100/min), pneumonia, ...) Clinical diagnosis is addressed in terms of a Table (99-3) on "Dichotomized clinical decision rule [*sic*] for suspected acute [PE]." The meaning here of "suspected" is left unspecified, and so is the operational meaning of, for example, "Alternative diagnosis less likely than [PE]." The seven items, when positive, are assigned numerical scores, and these are added up to obtain the total clinical-diagnostic score, in the range from 0 to 12.5. "Clinical probability of [PE] is unlikely with a score of 4 points or less; [it] is likely with a score of more than 4 points," with the (only) two probability values, "unlikely" and "likely," left unspecified.

This "decision-rule" dichotomy is an element also in diagnosis involving laboratory testing: Fig. 99-2 – "A CT scan-based algorithm for the diagnostic approach to suspected acute [PE]," with "suspected" still without operational meaning. The other elements are D-dimer test result, normal or abnormal; spiral CT scan result in terms of "PE present" or "inconclusive for PE" or "cannot be performed"; and leg ultrasound, in terms of "DVT present" or "DVT absent" – with all of these test results left unspecified. The algorithm leads to either "PE very unlikely – no treatment," or "Treat," or "VQ scan or pulmonary arteriogram." The text comments on the 'sensitivity' and 'specificity' of the D-dimer test, the spiral CT scan, and the VQ scan but, again, never defining the tests' positive/negative results.

After subsection on the "Treatment" comes one on "Prognosis." Its entire content is this:

> Most patients with PE who receive adequate anticoagulation survive. However, patients who are treated for PE are almost four times more likely (1.5% vs. 0.4%) to die from recurrent VTE in the next year than are those treated only for DVT. The 3-month mortality

is about 15 to 18%. In some series, PE itself has been the principal cause of death, whereas other series report that only 10% of deaths during the first year are attributable to PE. The presence of shock defines a three-fold to seven-fold increase in mortality: a majority of deaths appear to occur within the first hour of presentation. A potential long-term sequela from acute DVT is chronic leg pain and swelling (postphlebitic syndrome), which may result in significant morbidity (Chapter 81).

All in all, thus, Cecil displays, unwittingly, a remarkable inadequacy of the existing knowledge-base of gnostic probability-setting in eminent aspects of medicine – just as Davidson's does, only in quite different ways. The question arises: Are the 'specialty' textbooks better?

3.2 Specialty Textbooks on the Needed Knowledge

With the focus in Sect. 3.1 above on two example topics from each of cardiological and pulmonological medicine, drawing from contemporary general textbooks of medicine, instructive supplements to those descriptions-cum-annotations are their counterparts in textbooks specific to those two disciplines of medicine. I examine these two:

1. Fuster V, Walsh RA, Harrington RA (eds) (2011) Hurst's The Heart, 13th edn. McGraw-Hill Companies, New York
2. Mason RJ, Broddus VC, Martin TR et al (eds) (2010) Murray and Nadel's Textbook of Respiratory Medicine, 5th edn. Saunders/Esevier, Philadelphia

Hurst's first deals with *angina pectoris* in a Chap. 54 on "Coronary blood flow and myocardial ischemia," by C. Depre, S. F. Vatner, and G. J. Gross. Distinctions are made among secondary or effort-induced angina, primary angina or vasospasm, unstable angina, and angina arising from microvascular dysfunction. Following the descriptions of these comes the statement that "The primary manifestation of myocardial ischemia is angina pectoris, irrespective of the type of angina and the causal factors involved," and a description of this symptomatic manifestation of myocardial ischemia follows. But, very confusingly, this description is under the rubric of "Symptoms of angina pectoris," as though angina pectoris actually were not a symptom (of myocardial ischemia) but a type of disease (à la Osler, Sect. 3.1 above) – a type of "Coronary heart disease," which is the heading for the Part (8) in which this chapter is one out of 17.

A related other Chap. (56), by M. C. Kim, A. S. Kini, and V. Fuster, is entitled "Definitions of acute coronary syndromes" (plural; cf. Sect. 3.1 above). In the second paragraph of its two-paragraph introduction, to coronary heart disease, is the major orientational point that "CHD has been classified as chronic CHD, acute coronary syndromes and sudden death." The authors make no point of the lack of logic in this classification: logically, as I noted in Sect. 3.1 above, chronic CHD should be understood to have *acute CHD* as its alternative, the only one; and I here have to add that sudden cardiac death is not a type of CHD but a possible outcome of acute CHD.

"*Acute coronary syndrome* (ACS)," the authors say under the heading of "Acute coronary syndromes" (plural), "is a unifying term representing a common end result [*sic*], acute myocardial ischemia." The text that follows implies that under this singular term there actually are three acute coronary syndromes (each having myocardial ischemia as the "end result"): unstable angina, non-ST-segment elevation myocardial infarction, and ST-segment elevation myocardial infarction. But, while "unstable angina," meaning acute, partial occlusion of a coronary artery or impending MI (cf. Sect. 3.1 above), does have myocardial ischemia as its result, it does not have this ischemia as its "end result." And while myocardial infarction does have an "end result" – an outcome – this is either death or a sequela in the form of myocardial fibrosis, not myocardial ischemia.

A section on "Definition of unstable angina" follows, with this the first paragraph:

> Unstable angina is usually secondary to reduced myocardial perfusion resulting from coronary artery artherothrombosis. In this event, however, the nonocclusive thrombus that developed on a disrupted atherosclerotic plaque does not result in any biochemical evidence of myocardial necrosis. Unstable angina and [non-ST-segment elevation MI] can be viewed as very closely related clinical conditions with similar presentations but of different severity.

But what this really means is that "unstable angina" is a misnomer for acute CHD in the meaning of nonocclusive but ischemia-causing coronary thrombosis – distinct from acute CHD in the meaning of occlusive coronary thrombosis, which leads not only to myocardial ischemia but to myocardial infarction, subclassified according to absence/presence of ST elevation (though more logically according to the pathological duality signified by this). These three genuine subtypes of acute CHD do not have 'unstable angina' as a member but only as a common clinical (symptomatic) manifestation – in the usual meaning of acutely new or acutely worsened angina pectoris. (Cf. Sect. 3.1 above.)

The ensuing section in this chapter on "Definitions of acute coronary syndromes" is entitled "Non-ST-segment elevation myocardial infarction." It, however, does not define the nature – pathological – of this subtype of MI; it merely sketches its manifestations in ECG and chemical markers of myocardial injury. (Cf. Sect. 3.1 above.)

There are separate chapters for "Unstable angina and non-ST-segment elevation myocardial infarction" (Chap. 59) and "ST-segment elevation myocardial infarction" (Chap. 60), by different sets of authors. As before (Sect. 3.1 above), I focus on the former topic, the chapter by J. A. deLemos, R. A. O'Bourke, and R. A. Harrington, again following this by a brief commentary on the latter chapter.

In the section on "Diagnosis," there is this paragraph in the early subsection on "History":

> Evaluation of patients with suspected UA/NSTEMI should include the physician's opinion of the *probability* of the symptoms being caused by myocardial ischemia, categorizing the presentation into high-, intermediate-, or low-probability categories (Table 59-2). [Italics in the original.]

Actually, though, the "evaluation" – diagnostic work-up – of a patient with a presentation raising the possibility of acute myocardial ischemia (as one of the possibilities in the differential-diagnostic set, implied by the presentation) should result in the doctor's *knowing* (*sic*) about the probability that the sickness is caused by myocardial ischemia – and also in the counterparts of this for the other possibilities. That Table is about "Likelihood that chest symptoms [unspecified] are caused by myocardial ischemia attributable to obstructive coronary artery disease." Under "High likelihood" (unspecified) it lists "Coronary artery disease (particularly recent PCI)" and five other features, but without saying whether meant is any, or all, of the six features. Under "Intermediate likelihood" (also unspecified) the stated criteria are "Absence of high-likelihood features [all of them?] and any of the following," an example of the six items that follow being "Male gender."

Related to this is a later subsection on "Chest pain units," regarding which the orientational point is this:

> The evaluation of patients with chest pain who may have UA or MI is often difficult and uncertain. Hospitalizing all such patients for an extensive work up is unwise and results in unnecessary tests with patient risk and expense. However, missing the diagnosis of MI is the leading cause of malpractice claims for US ED [emergency department] physicians.

So, while diagnosis about acute CHD – its presence/absence – is often difficult yet critically important to achieve both swiftly and competently, an eminent textbook of cardiology fails to provide the needed knowledge-base for this. It does not even specify the patient presentation – as to chief complaint(s) of a person from a particular, broad demographic domain – that should raise the possibility and suspicion of acute CHD. It does not say what kind of chest pain is relevant as the chief complaint, nor whether dyspnea, for example, could be the chief complaint instead of chest pain. The items of history it presents (in that Table 59-2) as relevant for the diagnostic profile in the triage stage are remarkably superficial/nonspecific; and they reflect thinking about acute CHD without any regard for the differential-diagnostic set implied by the presentation, which set is left unspecified when addressing (in Table 59-2) the probability of acute CHD as a particular member of this set.

As the foregoing illustrates, such is, still, the knowledge-base of the clinical diagnosis about acute myocardial ischemia, that there can be no realistic expectation of more-or-less convergent diagnostic probabilities across even expert cardiologists experienced in emergency department work, when attending to cases of the patient presentation(s) suggesting, of all things, the possibility of a case of acute CHD.

Nothing at all is said about etiognosis, even though under "History" the second sentence points out that: "Cocaine use can cause coronary vasospasm and thrombosis in addition to its direct effect on altering myocardial oxygen demands through increases in heart rate and blood pressure; it has been implicated as a cause of ACS [ref.]," and even though etiognosis about cocaine use would have definite implications for the management of a case of acute CHD.

Under "Prognosis" are three brief paragraphs, none of them containing knowledge about prognostic probabilities. Available evidence on prospective course of health in cases of the subtypes of acute CHD is given only in the last one of those paragraphs, and it is as simplistic as this:

Data from the GRACE study show a 6-month mortality rate of 6.2% in patients with NSTEMI and 3.6% in those with UA. Rehospitalization rates over the 6 month period were approximately 20% and revascularization rates approximately 15% [ref.].

There is no attention to prognostic indicators nor to the nature of prophylactic treatments.

The confusion and superficiality displayed in this Chap. 59 on "Unstable angina and non-ST-segment elevation myocardial infarction" continues in the ensuing Chap. 60 on "*ST-segment elevation myocardial infarction*," by E. E. Hass, E. H. Young, B. J. Gersh, and R. A. O'Bourke.

The first section in this chapter – in the same "Part 8: Coronary heart disease" – is entitled "Epidemiology." The first one of its two paragraphs is about "coronary artery [*sic*] disease" (undefined) in its first sentence, "ischemic heart disease" (undefined) in the second sentence. The third sentence is about "acute myocardial infarction" (with chronic MI a non-topic in this Part 8 of the book, just as everywhere else); and it asserts that one-third of the cases of AMI in the U.S. are "caused by [*sic*] an acute [*sic*] ST-segment elevation myocardial infarction (STEMI) [ref.]." The rest of the paragraph is about AMI unspecified. The second one of the two paragraphs under "Epidemiology" is not about epidemiology in any existing meaning of this term; it is about the management of patients with STEMI. (Cf. "Angina pectoris" as Chap. 70 of *Cecil*, per Sect. 3.1 above.)

Just as in the previous chapter on UA and NSTEMI there is no definition of the nature (pathological) of the MI in the absence of ST elevation, so in this chapter on STEMI there is no word about the type of MI that manifests in ST elevation.

Under "Diagnosis" there are two paragraphs under the subheading of "Symptoms" and another two under "Physical examination," but nothing about diagnostic probabilities – even though under "Conclusion" said is that "The key steps in the management of STEMI patients include rapid diagnosis, ..."

There is no section on the knowledge-base of prognosis per se, though in the context of various treatments some evidence of their respective effects is superficially addressed. For example, under "Anticoagulation" the core content is this: In a particular, specified trial, "patients treated with enoxaparin plus tenecteplase had a lower combined end point of 30-day mortality, in-hospital reinfarction, and in-hospital refractory ischemic than did those treated with unfractionized heparin plus tenecteplase (11.4% vs. 15.4%; P = .0002)"; and in a specified other trial, "The combined primary end point of death or recurrent myocardial infarction at 30 days occurred in 9.9% of patients in the enoxaparin group and in 12% of those in the heparin group (P < .0001)." (The authors have an unstable conception of "end point" in intervention-prognostic research, here vacillating between mortality and death.)

Having thus sampled the contents of this book, it is of interest to glance at its Preface. It opens with this self-description:

> It has been 44 years since the publication of the first edition of *Hurst's The Heart*, the first multidisciplinary and comprehensive textbook on cardiovascular disease. Through 13 editions, *The Heart* has always represented *a cornerstone of current scholarship* [*sic*] in the discipline. Cardiologists, internists, and trainees from around the world have relied on its *authority*, breath of coverage, and *clinical relevance* to keep up to date with advances in the field and to help optimize patient care. [Some italics added.]

Turning to *Murray and Nadel's* (ref. above), it addresses *pneumonia* in its Chap. 32 on "Pyogenic bacterial pneumonia and lung abscess," by A. Torres, R. Menéndez, and R. Wunderink.

Without anywhere defining pneumonia, the authors say in the chapter's "Introduction" that "A clinical diagnosis of pneumonia can usually be readily established on the basis of signs, symptoms, and chest radiographs, although distinguishing CAP [community-acquired pneumonia] from conditions such as congestive heart failure, pulmonary embolism, and chemical pneumonia from aspiration is sometimes difficult." Then, under "Clinical presentation" said is the opposite: that "Unfortunately, information obtained from clinical history and physical examination is not sufficient to confirm the diagnosis of pneumonia. A definitive diagnosis requires the presence of new opacity on the chest radiograph."

Under "Patient evaluation" the first subsection addresses "Clinical evaluation." In it the clinical-diagnosis probability of CAP is addressed only by the first two sentences: "The clinical findings that best differentiate CAP from other acute respiratory tract infections are cough, fever, tachycardia, and pulmonary crackles [ref.]. CAP is present in 20% to 50% of persons who have all four factors [ref.]." The shared chief complaint is left unspecified in this context, and so is the differential-diagnostic set.

The subsection on "Radiographic evaluation" opens with what was already alluded to in the "Introduction": "Radiographic evaluation is necessary to establish the presence of pneumonia, ... [refs.]." But the same paragraph points out that "Frequently, the emergency department admissions with radiographic pneumonia are subsequently proved not to have pneumonia." The pattern of "radiographic pneumonia" is left undefined. However, "Failure to detect abnormal opacity on the plain chest radiograph essentially rules out pneumonia [ref.]." This "evaluation" is not addressed as a supplementary test adding to the diagnostic profile but, instead, as self-contained diagnostic process.

A separate, brief subsection on "Differential diagnosis" has a table listing "Noninfectious causes of fever and radiographic changes that mimic community-acquired pneumonia." That subsection, too, is devoid of specification of the context of this in terms of the patient presentation and, also, of any content on the diagnostic probabilities of pneumonia and of those pneumonia-mimicking conditions.

Related to prognosis, under "Therapeutic approach to pneumonia" the first subsection, on "Assessment of severity," addresses two scoring schemes to assess the risk of fatal outcome of CAP with a view to the choice between outpatient treatment and hospitalization. Then, following an extensive coverage of treatments, there is a subsection on "Clinical course" separately under pneumococcal and other types of CHP. In respect to the pneumococcal type of CAP the essential content of this presentation – in one very brief paragraph – is this: "With an appropriate antibiotic, a salutary clinical response usually occurs within 24 to 48 hours. . . . The overall mortality of bacteremic pneumonia is 11% to 20% [ref.], increasing to 20% to 40% in persons over 65 years of age."

The Chap. (51) on *"Pulmonary thromboembolism,"* by T. A. Morris and P. F. Fadullo, begins (under "Introduction") with these highly discouraging words:

> The one generalization about venous thromboembolism (VTE) that is free from controversy is that many aspects of this disorder remain controversial. There are multiple reasons why VTE continues to engender lively debate. Perhaps the major reason, . . ., is that a number of fundamental questions continue to exist regarding the pathogenesis, clinical presentation, diagnosis, and therapy of the disease.
> VTE represents a potentially fatal disease process with a clinical presentation that is often silent or nonspecific and for which a wide range of diagnostic techniques is available, many with technical and interpretive limitations. . . .

The chapter addresses diagnosis under the heading "Clinical prediction rules," even though prediction is not in the essence of diagnosis. The presentation is essentially the same as in the general textbook of *Cecil Medicine* (Sect. 3.1. above).

The knowledge-base of prognosis is not a topic as such, nor even tangentially under the treatment of PE (by heparin).

All in all, thus, the inescapable impression is that, even in textbooks for particular disciplines of medicine, the proffered knowledge-base of medicine remains highly deficient in form; and in the content of whatever form, confusion and contradictions, including self-contradictions – even about the substantive concepts – are prevalent.

3.3   Medical Journals on the Needed Knowledge

Textbooks of medicine draw their directly practice-oriented – quintessentially applied – scientific contents from publications in 'medical' journals, which remain undifferentiated according to the needs of practitioners on one side and medical scientists on the other. From such journals, a practitioner would naturally look to find advances in the gnostic knowledge-base of his/her particular discipline of medicine (Sects. 1.1 and 1.2), while for a scientist of the meta-epidemiological clinical sort (Preface) the natural concern would be to find new evidence about gnostic probabilities (Sects. 1.1 and 1.2) and commentaries on the available evidence per se as well as on its inferential implications for the advancement of gnostic knowledge.

A practitioners-oriented medical journal can rationally be directed to practitioners in general, regardless of their particular disciplines of medicine – naturally addressing general concepts and principles of medicine and knowledge on substantive topics of general relevance in medicine, such as the prognostic implications of the use of analgesic, anti-inflammatory, and antibiotic medications, but not topics such as bariatric or cardiac surgery, for example. But it is unnatural of a journal to be oblivious to the differences between the respective needs of practitioners and researchers, and to the differences in needs among the different disciplines of medicine.

As an example of how a 'medical' journal now unnaturally endeavors to serve both practitioners and researchers, and both of these in all disciplines of medicine, I here examine *JAMA, The Journal of the American Medical Association*, the 24 issues of it that constitute its volume 307, covering the period of January-June 2012. This is the journal of choice here in part because the AMA was the instigator of the highly influential report that a century ago introduced one of the two now-prevailing misconceptions about the essence of scientific medicine (Sect. 4.1).

In the *number 1* issue of this volume, an early page is devoted to specification of the Journal's "Key objective" and its "Critical objectives." The *key objective* is said to be, "To promote the science and art of medicine and the betterment of the public health." The implication of this appears to be that, in this Journal, "the science and art of medicine" continues to be viewed as a unitary entity, despite the medicine versus medical science duality (Sect. 4.1) and the differentiation of modern medicine into its particular disciplines; that the Journal provides content all of which is relevant for all physicians without regard for these distinctions. Indeed, a full-page advertisement in earlier issues of the Journal has been saying that *JAMA* – implicitly all of it – is "A must see and read for all medical professionals."

Among the *critical objectives*, the first one out of a total of 10 is, "To maintain the highest standards of editorial integrity independent of any special interests." Immediately notable in respect to this is, however, that the first two pages of this issue of the Journal are devoted to an advertisement by a pharmaceutical company, and so is the back cover.

The second one of these objectives is, "To publish original, important, valid, peer-reviewed articles on a diverse range of medical topics." One implication of this is that, as for the knowledge-base of scientific medicine, published is only evidence bearing on it, and only original (rather than also derivative, review-type) evidence, without any new knowledge being published. And another implication is that the topics of research in the Journal are diverse because all important topics of research for medicine are of professional concern to all physicians, regardless of their particular disciplines of medicine. This implication indeed is explicit in Critical Objective number five: "To enable physicians to remain informed on multiple areas of medicine, including developments in fields other than their own." But it remains a mystery to me why bariatric (obesity-reducing) surgery, for example, is seen to be of professional concern for an ophthalmologist and a dermatologist more than it is of such concern for a computer engineer or a professional athlete, say.

Turning to the issue-specific contents in that volume of *JAMA*, I initially focus comprehensively on this number 1 issue, on the contents of this concerning the knowledge-base of medicine and the evidentiary precursors of this.

There first are two articles under the rubric of *"Medical news & perspectives,"* one of them, by M. Mitka, a depiction of the prevailing confusion about the intended effects of medicational interventions to increase the serum level of high-density lipoprotein cholesterol. The other one, a "Viewpoint" article by V. Prasad, A. Cifu, and J. P. A. Ioannidis, points out, for example, that "Medical practice has evolved out of centuries of theorizing, personal experiences, bits of evidence, expert consensus, and diverse conflicts and biases. Rigorous questioning of long-established practices is difficult. There are thousands of clinical trials, but most deal with trivialities or efforts to buttress the sales of specific products." But: isn't this known to be true rather than simply the authors' viewpoint?

Then, after these sad up-front articles there are three others under *"Original Contribution"* (singular): "Effect of dietary protein content on weight gain, energy expenditure, and body composition during overeating. A randomized trial," by G. A. Bray, S. R. Smith, L. de Jonge, et alii; "Bariatric surgery and long-term cardiovascular events," by L. Sjöström, M. Peltonen, P. Jacobson, et alii; and a third one with no bearing on the scientific knowledge-base of medicine. The three have almost the same structure for the up-front abstract: Context; Objective; Design, setting and participants; Intervention; Main outcome measures; Results; and Conclusions – except that Intervention is conspicuously missing from one of the articles on the effects of an intervention (bariatric surgery). Remarkable about this routine structure of the abstracts, most superficially already, is that its section-headings do not concur with those in the actual body of any of the articles.

The Conclusions from the first one of these studies were these: "Among people living in a controlled setting, calories alone account for the increase in fat; protein affected energy expenditure and storage of lean body mass, but not body fat storage." According to the ninth Critical Objective of the Journal, this study and its Conclusions exemplify the Journal's efforts "To achieve the highest level of ethical medical journalism and to produce a publication that is timely, credible, and enjoyable to read."

As for the ethics of this study as such, distinct from the report on it, its protocol was "approved by [its local] institutional review board," and "Participants provided written informed consent." The Journal presumably had no role in this other than verifying that these statements were included in the report – in reference to the local ethics represented by the local IRB. Thus the Journal presumably did not review the local IRB report to determine whether the study's approval was consistent with national or international standards of ethics in medical research, implicitly holding the highly questionable view that the ethics of medical research is a local matter, fairly represented by the decisions of local IRBs. In respect to whatever standards of research ethics, the Journal presumably subscribes to the view – also questionable – that reports of unethical studies, however valuable scientifically, should not be published; that publishing them would be unethical. I remain unaware of any serious attempts to justify this, and I wonder: Is the empirical knowledge-base of criminology unethical to the point of being downright illegal?

This study report satisfied the concern for timeliness in the sense that obesity now is a very topical issue as such and specifically in respect to its dietary etiology – which, remarkably, remains a matter of controversy. But the timely issue in this, as for the type of diet conditionally on its energy content, is not its relative proportions of fats and proteins in the context of a given level of carbohydrate intake; the timely issue is the relative proportions of fats and carbohydrates (refs. below).

References:

1. Taubes G (2011) Why we get fat an what to do about it. Alfred A. Knopf, New York
2. Taubes G (2007) Good calories, bad calories. In: Fats, carbs, and the controversial science of diet and health. Anchor Books, New York

As for the credibility of this report, a distinction needs to be made between the evidence in it and the Conclusions from this evidence. The reported evidence a reader must take at face value, except when, later and very exceptionally, reasons to suspect fraudulent reporting emerge. But as for the Conclusions from the evidence, as they are presented in this study report or in the report on any single original study alone, well, producing and reporting them is scientific malpractice, 'medical' journals' common demands for these notwithstanding; and this is particularly so when the authors' own, inherently biased opinions about the truths in question are the meanings of these Conclusions. All of the Conclusions in these journals are devoid of credibility, notably when they indeed are of the form of conclusions instead of being mere restatements of the results.

Insofar as conclusions nevertheless are drawn from the evidence of a particular hypothesis-testing study alone, they need to be consistent with the nature of the hypothesis and of the evidence, to say nothing about them having to be of the form of scientific propositions (the veracity of which is being claimed). To conclude that "Among people living in a controlled setting, calories alone account for the increase in fat" was inconsistent with the hypothesis, which had no specificity or reference to "people living in controlled circumstances" (such as those of prisons). And this conclusion was inconsistent with the nature of the evidence, which had nothing to do with whether the diet's composition in terms of its carbohydrate content also accounts for increase in adipose tissues. And to say that "protein affected . . ." is not even of the form of scientific propositions and conclusions/knowledge; for these have no particularistic referents (such as a particular study) but are abstract, devoid of spatio-temporal coordinates, and they therefore are expressed in present tense, not past tense.

All in all, thus, this article is not "enjoyable to read" to anyone familiar with the standards of scientific thought and writing, respects these, and expects them to prevail in reports on medical research.

From that second original study the authors' Conclusions were these: "Compared with usual care, bariatric surgery was associated with reduced number of cardiovascular deaths and lower incidence of cardiovascular events in obese adults." But: good-quality research does not deal with ill-defined, nonspecific entities such as "usual care"; it does not address an intervention, such as bariatric surgery, in respect to its "associations" with subsequent phenomena of health but addresses, instead, its effects on these (when used in lieu of an expressly defined alternative intervention). In good reporting of the evidence from an intervention-prognostic study, a lower number is not equated with "reduced" number; deaths are not addressed as their number per se, and notably when the numbers of other events are addressed as the rates of incidence they imply; and in particular, insofar as there is, contrary to principle (above), reporting of Conclusions, the statements of these should not be of the form summaries of findings/results (past tense) but of scientific knowledge (abstract, expressed in present tense; cf. above).

The Conclusions from the remaining third Original Contribution – by R. D. Kociol, R. D. Lopez, and R. Clare, et alii – were these: "In this multinational study, there was variation across the countries in 30-day readmission rates after STEMI [ST-segment elevation myocardial infarction], with readmission rates higher in the United States than in other countries. However, this difference was greatly attenuated after adjustment for length of stay." Again, thus, a summary of findings/results is presented as Conclusions. But as I noted above, at issue in this study is not the (abstract) knowledge-base of the practice of medicine (but, instead, particularistic practices themselves).

This article's final section is entitled "Limitations," and it opens by saying that "Our study has several limitations." But there are no studies free of several limitations, while this particular study has one unusually conspicuous limitation, which the authors do not mention: Its scores of P-values and 'odds ratios' not only do not serve to advance the knowledge-base of cardiological practice; they don't serve any other scholarly purpose either.

A section entitled "*Clinician's Corner*" follows, the title quite remarkably suggesting that only this section among the substantial number of them is devoted to clinicians; and in it there are only two articles: "Management of needlestick injuries: a house officer who has a needlestick," by D. K. Henderson, and "Update: a 52-year-old woman with disabling peripheral neuropathy: review of diabetic polyneuropathy," by A. N. Ship and N. S. Trivedi. Said in the up-front abstract of the former is that "Using the case of a house officer who has a needlestick during a resuscitation attempt, prevention of needlesticks including universal precautions and post exposure management of occupational HIV, hepatitis B, and hepatitis C exposures is discussed." This article indeed is a mix (curious) of a case with a review of the literature, the latter extensive to the point of involving as many as 99 articles.

The other piece involves a very brief case report on a patient – she's a nurse – culminating in her statement that, "I wish that someone would talk to people with diabetes before things become this severe. I wish I'd known," implying that doctors of diabetics don't do this; but she presents no evidence to justify this lament, nor do the authors. The review of the literature on diabetic polyneuropathy is unrelated to this essential point of the case report.

Then come two "*Editorials*," one on each of the first two Original Contributions. The title of the first one, by Z. Li and D. Heber, declares that "Extra calories increase fat mass while protein increases lean mass." Evidently taking, unjustifiably, the study authors' Conclusions to represent new knowledge, the editorialists fail, in this declaration, to fairly present them, presumably in part because those Conclusions were not well written (cf. above). In truth the study merely suggests that introduction of extra calories from, specifically, proteins and fats increases fat mass independently of the respective proportions of these, and this, specifically, when the consumption of carbohydrates remains constant (and not excessive); and the study suggests that, in that restricted context, lean body mass is increased by dietary proteins but not by dietary fats. The editorialists unjustifiably say that the study authors "found" the corresponding truths; that they "showed" the hypotheses to be true.

These editorialists also say that the study in question "informs primary care physicians and policy makers about the benefits of protein in weight management. The results suggest that . . ." So now the study no longer is presented as providing for categorical conclusions but merely as suggestive of the truths in question, which it indeed should be seen to be (cf. above); and it is presented as informing primary-care physicians. But insofar as it indeed informs primary-care physicians, it equally informs community-level preventive medicine, epidemiologists' health education in this, while its relevance to 'policy makers' is not at all obvious.

Relating to the bariatric surgery study by Sjötröm et alii, the other Editorial, by E. H. Livingston, focuses on "Inadequacy of BMI as an indicator for bariatric surgery." The leitmotif is that "Interpreting the clinical importance of these results [of Sjöström et alii] is complicated by the authors' findings that reduced cardiovascular events among the surgical patients were not related to either baseline weight or weight loss." In line with those results, this Editorial points out that, as previously reported from that same trial, "patients undergoing bariatric surgery experienced a 30% reduction [*sic*] in all-cause mortality [ref.]." And it refers to another study in which "the authors found 40% reduction in all-cause mortality [ref.]." Then, from the premise that "the benefits from bariatric surgery operations are not related to weight loss" coupled with the added ones that "body weight and BMI do not reflect body fat accumulation" – these added ones left unjustified – this Editorial deduces its main point, namely that "BMI alone should not be used as a criterion for obesity treatment by bariatric surgery operations."

So again, in this Editorial, the observation of a different rate subsequent to an intervention, when compared with its alternative, is equated with observation of the intervention's effects – here a "reduction" produced by the intervention – as though effects of a cause were phenomena and thereby observable. Kant taught us that causal connections are 'conceptions a priori' – noumena – rather than concepts arising from experience with phenomena. And my *The American Heritage Dictionary of the English Language* affirms my understanding that 'reduction' is not a synonym for 'decline'; that it is "The act or process of reducing," as by a preventive intervention.

It is not clear to me why an Editorial should bring out references to the results of other studies when these already were addressed in the original contribution in question. And it is particularly unclear to me why an Editorial would equate the results and the (unjustified) Conclusions from an original contribution with knowledge and then go beyond, to invoke another piece of purported knowledge, and 'deduce' a *guideline* for practice from these. Even if these predicates were correct, their practice implications do not rationally flow from them alone, as a succession of scholars have long understood and eloquently pointed out (see, e.g., App. 1 in ref. 3 given in Preface).

And the related larger question is this: Why these Editorials? why not the alternative of simply inviting commentaries from members of the relevant scientific community out there? And the related, rather disturbing question is this: Are the Editorials part of the Editors' way of controlling the discussion about the evidence they (elect to) publish and the perceived knowledge implications of this? Particularly suggestive of this unwholesome possibility is the fact that there appears to be the policy of having only one Editorial concerned with any given original contribution, and an invited one at that.

I address the Sjöström et alii study in Sect. 7.5.

The last section in this issue 1 of volume 307 of *JAMA* is *"Online CME quiz questions."* In this particular issue the heading of these continuing medical education questions defines the topic to be "Management of needlestick injuries: a house officer who has a needlestick," thus harkening back to the first one of the two Clinician's Corner articles in this issue of the Journal. Presented are five questions together with four possible answers to each, one of the questions having to do with what the practices are of "Surgeons who use the 'no-touch' technique." There is no specification of the disciplines of medicine whose CME is at issue, the implication being that psychiatrists and diagnostic radiologists, for example, can demonstrate their continuing education in their respective disciplines by their having gone to identify the 'correct' one among the presented possible answers to these simplistic questions from readily available source, this on a topic of no consequence in their particular disciplines of medicine.

Turning to the *number 2* issue of that recent volume of *JAMA*, a notable generic novelty relative to the first issue is the first one of the two Clinician's Corner articles. This article, by L. C. Yourman, S. J. Lee, M. A. Shonberg, et alii is entitled "Prognostic indices for older adults: a systematic review." Associated with it is an Editorial, by T. M. Gill, entitled "The central role of prognosis in clinical decision making," and the Yourman et alii article also is the topic of the Online CME Quiz Questions. I here focus on that review article and the Editorial related to it, without considering the rest of this issue of the Journal, thus focusing on what is new in kind – including the publication of also derivative (review-type) articles and not only original studies (contrary to what was said the Journal's second Critical Objective in issue number 1 of this volume).

An *original study* I take to be one that produces new evidence, whether from the very first study on the object of study at issue or from a replication of this study. Such a study I distinguish from a *derivative study*, in which evidence from some original studies on the same object of study is synthesized to derive a result of greater precision. A derivative study naturally involves a review of the original studies on the object of study at issue. A *good* derivative study has a number of virtues, being systematic – protocol driven – representing but one of these. It thus remains a mystery to me why derivative studies in *JAMA*, as elsewhere, are routinely advertised as being based on a 'systematic [*sic*] review.'

That review by Yourman et alii had as its Objective, as stated in the abstract, this: "To assess the quality and limitations of prognostic indices for mortality in older adults through systematic review." So, theirs actually was not a report of a 'systematic review' in the usual, derivative-study meaning of this term. Their actual objective evidently was to evaluate existing 'indices' – prognostic probability functions, one should presume – for 'mortality,' meaning death not otherwise specified; that is, to critically discuss the qualities of the studies behind the various probability functions. I need to point out that mortality is a term and concept of community medicine, not clinical medicine; mortality is the 'emergent' quality of frequency of occurrence that death has in population contexts in contrast to the individual contexts of clinical medicine. A patient's death is not a case of 'mortality' in the terminology of clinical medicine.

Regarding the objective of that project, a somewhat more elaborate specification of it is given in the last paragraph of the article's introduction:

> We performed a systematic review to describe [*sic;* cf. above] the quality and limitations of validated non-disease-specific prognostic indices that predict [*sic*] absolute risk of all-cause mortality [see above] in older adults. Recognizing that older adults are more likely to have more than 1 chronic illness than younger adults, we focused on non-disease-specific indices.

But it still remains very unclear what the authors took "prognostic indices" to be in general, and what their generic "non-disease-specific" subtypes to them were in particular. As for risk of death (not "mortality," cf. above), this is not a topic of practice for "prediction" but for assessment and ultimately knowing, given the ascertained prognostic profile of the person at the time. And that profile naturally must account for the survival-relevant illnesses the person is known to have at the time, while the particular cause of death may not be of material concern in the prognosis when decision about treatment is not at issue; the outcome being addressed thus can be illness non-specific. But now the question arises whether the objective indeed was to "assess," de novo and critically, the quality and limitations of these studies or merely to "describe" their already known nature in these respects.

Might the Conclusion in the abstract help to determine more closely what this project was about? Said under this heading in the abstract is the following:

> We identified several indices for predicting overall mortality in different patient groups; future studies need to independently test their accuracy in heterogeneous populations and their ability to improve clinical outcomes before their widespread use can be recommended.

So, nothing at all is either (summatively) said about or concluded from the original studies that were reviewed. Instead, some gratuitous and questionable advice is offered in respect to future studies; and in this, a major conceptual mistake is made by implying that (the deployment of) indices of risk of death ought to improve clinical outcomes, when such improvement properly is understood to be the intended consequence of interventions rather than of prognostications. In the absence of effective, course-changing intervention, prognosis serves to help make suitable adaptive arrangements in the light of the prognosis.

That pseudo-conclusion follows this Results section in the abstract:

We reviewed 21953 titles to identify 16 indices that predict risk of mortality [see above] from 6 months to 5 years for older adults in a variety of settings: . . . At least one measure of transportability (the index is accurate in more than 1 population) was tested for all but 3 indices. . . . Only 2 indices were independently validated . . .

To clinicians, reading their particular "corner" of the Journal, this passage in the article's abstract poses a major challenge for understanding.

The Methods section in the article proper reveals the source of the confusion, as the second one of its mere two paragraphs opens with this devastating point: "There are no accepted criteria to assess the quality of prognostic indices. Therefore, we adapted criteria from previous work published by experts in medicine and epidemiology [refs.]" – revealing that the authors had no established or otherwise solid theoretical framework for doing what was their objective to do, and that this was so because the theory of quintessentially applied medical research remains quite underdeveloped in this respect.

Despite the unavailability of accepted criteria for the assessments, the related *Editorial*, by T. M. Gill, declares that "The authors *rigorously assessed* each index for generalizability, accuracy, and potential bias [italics added]." From this misrepresentation of the Yourman et alii article the Editorial proceeds to the claim that "The focus on mortality [*sic*; see above] risk rather than life expectancy, . . . , limits the use of currently available prognostic indices" – as though the risk of death (n. o. s.) as a function of prospective time would not imply the median of the remaining life span (along with other centiles of the distribution of this).

The Editorial implies that something else also limits the use of currently available prognostic indices, and even more:

To determine whether use of a previously validated prognostic index is better than usual care, an impact study must be conducted [ref]. The preferred design is a randomized controlled trial, in which the effect of using the prognostic index is evaluated on physician behaviors and patient outcomes compared with a control that does not include the index.

This, I take it, is to say, for one, that it is not clear a priori that a doctor's knowing prognostic probabilities (on the basis of validated indices) is better than the usual care characterized by not knowing them. In other words, implied seems to be that it is not clear a priori that it is good in a doctor's practice that (s)he comes to know about the client's future health – as the basis for teaching the client about this (Sect. 1.1), and for the client's and/or the doctor's prospective decisions just the same.

Equally curious is the more specific implication that if doctors' knowing about their clients' health (incl. survival), replacing ignorance about this, does not improve doctors' behaviors and their clients' health, then the knowing is useless. Health cannot be expected to be improved by prognosis per se but only by effective intervention suggested by intervention-prognosis (cf. above), and intervention effects were not, as I noted above, at issue in the article by Yourman et alii.

As a closing note on this number 2 issue of the January-June 2012 volume of the Journal, I simply point out that there are three Original Contributions and that all of them have but a vague bearing on the advancement of the knowledge-base of medicine.

Notable for the purposes here in the *number 3* issue of the volume is again an article, now the only one, in the Clinician's Corner together with its associated Editorial, and also the report on a randomized trial the report on which is one of the two Original Contributions.

The article in the Clinician's Corner is entitled "Symptomatic in-hospital deep vein thrombosis and pulmonary embolism following hip and knee arthroplasty among patients receiving recommended prophylaxis: a systematic review," with J. M. Januel, G. Chen, C. Ruttieux, et alii the authors. According to the up-front abstract, the Objective was "To establish a literature-based estimate of symptomatic VTE [venous thromboembolism] event rates prior to hospital discharge in patients undergoing TPHA [total or partial hip arthroplasty] or TPKA [total or partial knee arthroplasty]." And the Conclusion was this: "Using current VTE prophylaxis, approximately 1 in 100 patients undergoing TPKA and approximately 1 in 200 patients undergoing TPHA develops symptomatic VTE prior to hospital discharge."

For the associated Editorial, by J. A. Hert, the title was "Estimating the incidence of symptomatic postoperative venous thromboembolism: the importance of perspective." This brief article emphasizes the importance of perspective in the meaning of the duration of the postoperative period of DVT risk extending well beyond that of hospitalization, but notes that this point actually was made by Januel et alii.

In these superficial terms, both the study (derivative) report and the Editorial are well understandable to whoever is concerned to deploy the reported rates as single-valued benchmarks for the evaluation of the quality of TPHA and/or TPKA care as for the prophylaxis of VTE, while prognosis about VTE naturally would require distinction-making according to prognostic indicators and also according to the choice of prophylactic intervention (Sect. 1.4). On the other hand, though, the full report by Januel et alii again is illustrative of the hypertrophy of statistics that has come to characterize articles in 'medical' journals even as it presumably is well beyond the comprehension of many of the researchers among their readers – and, generally, more a matter of affected mannerisms and unnecessary distraction than exposition of the genesis of the core results.

An example of the statistical overkill and obfuscation is, already, the second sentence under Statistical Analysis: "We used a fixed-effects model where appropriate, weighting each estimate by its standard error, using the Mantel-Haenszel method and a random-effects model according to the method of the DerSimonian and Laird [refs.]." Without any elaboration this was immediately followed by a statement of how the authors "evaluated heterogeneity," namely "using I^2 statistics [refs.]" in such a way that when $I^2 > 50\%$ "we considered intracluster homogeneity as not assessed," also left without elaboration.

Not only is this statistical writing remarkably obtuse, but its title and first sentence already are profoundly in error. At issue is not analysis but synthesis of the data; and, a parameter's empirical values from different studies are to be weighted according to their respective amounts of information, and these are not proportional to the respective SEs but to the inverses of the squares of these. And scarcely do the clinicians reading their "corner" know the distinction between a fixed-effects and random-effects model and the proper choice between these two in the situation at issue here.

The Original Contribution I alluded to was entitled "Bridging antiplatelet therapy with cangrelor in patients undergoing cardiac surgery: a randomized trial," the authors being D. J. Angiolillo, M. S. Firstenberg, M. J. Price, et alii for the BRIDGE Investigators. According to the abstract, the Objective was "To evaluate the use of cangrelor, an intravenous, reversible $P2Y_{12}$ platelet inhibitor for bridging thienopyridine-treated patients to coronary artery bypass grafting (CHBG) surgery." And the corresponding Conclusions (plural) was expressed by this singular statement: "Among patients who discontinue thienopyridine therapy prior to cardiac surgery, the use of cangrelor compared with placebo resulted [sic] in a higher rate of maintenance of platelet inhibition."

Rather than here merely continuing to express my problems with the writings of the Objectives and Conclusions in the Journal's articles, I now first make a positive observation about these writings: since there are no uncontrolled randomized trials, it is good to follow these authors' style and write, simply, about "randomized trials." But the problems continue to abound, starting from the statement in the abstract's "Design, setting, and patients" section. The types of setting and patient are not extrinsic to the design; they are topics in the process of the design and elements in the design arising from this process. To say that the designed study was characterized by being a "Prospective, randomized … trial" is to imply that a randomized trial could alternatively be retrospective. And the problems continue from the very first sentence of the text proper, which addresses "the standard of care to prevent [*sic*] the risk of recurrent atherothrombotic events in high risk settings, ..." Risk can in principle be reduced and even eliminated, and exposure to it can be avoided; but prevented by preventive intervention is not any risk but the health phenomenon (adverse) whose risk is at issue.

This trial's design – in which the selection of the study setting and the participating patients are component issues rather than extrinsic to it (cf. above) – ideally would be understood to define the study plan all the way to the production of its results, thus encompassing also the synthesis (*sic*; see above) of the acquired data. But as remains a routine in these reports (per a demand of the journals' Editors), there is a section entitled Statistical Analysis also in this trial's report; and it is, as usual, rather devoid of content that is specifically statistical, relevant, and correct – as well as understandable to the intended readers (cf. above). I'll comment on the routine first topic in this section (one that is there in randomized-trial reports by Editors' express demand).

The topic in question is 'sample-size calculation,' specified for this trial as follows:

> We estimated that, assuming 30% of placebo-treated patients and 60% of cangrelor-treated patients would reach the primary end point, a sample size of 106 patients (53 for each group) would provide 90% power to detect a statistically significant difference between the treatment groups at a two-sided α of .05 using a χ^2 test. ... We selected a sample size of up to 100 patients per treatment group to further evaluate the safety of cangrelor before and after CHBG surgery.

Whether, on that (rather arbitrary) "assumption," calculated was the corresponding "power" – probability of getting $P < \alpha$ – conditionally on the "sample" size having been set a priori, or the other way around, the presentation of the result of this calculation is irrelevant. A token of this irrelevance is that the enrollments into the trial continued till the P-value (presumably one-sided, given the one-sided hypothesis) was infinitesimally small, as the χ^2 (1 d.f.) got to be as large as 110. $P < 0.05$ was, thus, achieved when the study size was but a tiny fraction of what it got to be. Presentation of that calculation has no bearing on the evidentiary burden of the study result; only the result's genesis has. Nor was any relevance attributed to it, as is routinely the case.

Why the study's 'data safety monitoring board' did not call a halt to the enrollment of patients into the trial earlier, very much earlier, is not explained in the report's Statistical Analysis section, nor elsewhere in the report. And related to this, there remains the strong and troubling presumption that while many – the majority – of the patients were recruited when there already was "a statistically significant difference between the treatment groups," as defined (above), these later-enrolled patients were not informed about this statistical fact despite the critical bearing it would have had on their willingness to participate.

In the remaining 21 issues of the recent volume (307) of *JAMA* my concern was to focus on the first two articles with bearing on the knowledge-base of diagnosis and also the counterparts of these in respect to etiognosis about adverse effects of medication use, having already dealt with two articles bearing on prognosis – one with and one without concern for the effects of treatment.

In issue *number* 6, in its Clinician's Corner under the subrubric of "The rational clinical examination" is an article entitled "Does this patient have an infection of a chronic wound," authored by M. Reddy, S. S. Gill, W. Wu, et alii. In its abstract the Objectives are said to have been "To determine [*sic*] the accuracy [*sic*] of clinical symptoms and signs to diagnose infection in chronic wounds and to determine whether there is a preferred [*sic*] noninvasive method for culturing chronic wounds [*sic*]." To this end the authors reviewed original studies and came to the Conclusions that "The presence of increasing pain may [*sic*] make infection of a chronic wound more likely. Further evidence is required to determine which, if any, type of quantitative swab is most diagnostic." No Editorial is associated with this article.

Now, insofar as a study – in this instance a derivative, review-type study – is about symptoms and signs in the diagnosis of a particular illness, it should not be about the diagnostic "accuracy" of symptoms (which all are, by definition, "clinical") and signs; its proper objective should be understood to be the production of evidence about the magnitudes of the parameters in an appropriate object of study, in a well-designed diagnostic probability function for the presence of the illness in a suitably defined domain of the diagnostic concern (Sect. 2.1).

In the Reddy et alii study there wasn't even a hint about concern for such an object of study, with intent to focus on original studies addressing that function and the concern to synthesize their respective results. Instead,

> Articles were included if they were original studies describing historical features, physical examination maneuvers, or [*sic*] laboratory markers or [*sic*] were radiologic studies in the diagnosis of infection of chronic wounds among adult patients; if data could be extracted to construct 2 × 2 tables or the study reported operating characteristics of the diagnostic measure; and if the diagnostic test [*sic*] was compared with the reference standard test . . . or other commonly used standards . . .

The included studies were subjected to Quality Review by three of the authors:

> Each study was rated with a topic-specific quality rating scale that used published principles [ref.], as well as a modified quality checklist specific to the Rational Clinical Examination series (BOX) [ref.].

That box is about "Criteria for level of evidence in diagnostic studies." There are five of these levels, starting with:

> Level 1: independent, blind comparison of test (i.e., sign, symptom, or investigation) results with reference standard of anatomy, physiology, diagnosis, or prognosis among a large number of consecutive patients ($\geq$75).

Further in line with the prevailing culture of diagnostic research, the authors were concerned, first, with "prior probability" of infection in the wounds, meaning probability in total disregard of the particulars of the wound:

> The prior probability of infection in a chronic wound can be inferred from the prevalence of infection of chronic wounds in the level 1 to 3 studies . . . the level 4 to 5 studies . . . in all 15 studies . . .

And the remaining concern was with "odds ratios" as measures of the "accuracy" of the various "tests."

As another manifestation of this prevailing culture of diagnostic research, in *number 8* – also in Clinician's Corner and under the subrubric of "The rational clinical examination" – there is an article entitled "Does this patient with liver disease have cirrhosis?" authored by J. A. Udell, C. S. Wang, J. Tinmouth, et alii. Very similar to the review study by Reddy et alii addressed above, this study, too, was left without an Editorial. Its main Conclusion was this: "For identifying cirrhosis, the presence of a variety of clinical findings or abnormalities in a combination of simple laboratory tests that reflect the underlying pathophysiology increase its likelihood" – just as simplistic and unsurprising (i.e., meaningless) as the Conclusion by Reddy et alii that "The presence of increasing pain may make infection of a chronic wound more likely."

It deserves note that both of these diagnostic-study reports appeared under a rubric implying that at issue was *rational diagnosis* and, presumably, rational diagnostic research. Both the practice and the research remain remarkably deficient in their rationality of theory – concepts and principles – as I explain in Sect. 2.1 and Chap. 5 of this book. From the Reddy et alii study one learns nothing about setting the probability of infection for a chronic wound; and from the Udell et alii study the reader learns nothing about the way in which diagnosis about cirrhosis of the liver can be achieved in a patient with liver disease. These reports should have been associated with Editorials – proclaiming their meaninglessness.

In issue *number 13*, under Original Contribution, there is an article entitled "Oral fluoroquinolones and the risk of retinal detachment," authored by M. Etminan, F. Foroogian, J. M. Brophy, et alii. According to the abstract, the Objective was "To examine the association between use of oral fluoroquinolones and the risk of developing retinal detachment," and the Conclusion was this: "Patients taking oral fluoroquinolones were [*sic*] at a higher risk of developing retinal detachment compared with nonusers, although the absolute risk of this condition was [*sic*] small."

For orientation here, the need is to understand whether this was, as the stated concern for "risk" suggests, a prognostic study. If this was the case, the study was addressing the risk of future occurrence of retinal detachment as a joint function of prospective use/non-use of the medication, causally, and prognostic time at least, if not also of non-treatment determinants of the risk, acausally – this prospective function from the vantage of a defined domain for the prognostication. The quotations above do suggest that this was the case, as "risk" is so eminent in them (cf. Sect. 1.5).

The other possibility is that this was an etiognostic study, intended to serve causal explanation of instances of the adverse event (retinal detachment) in a defined domain of their occurrence, specifically the probability that the event's occurrence in association with the medication's antecedent use is caused by that antecedent. To this end studied would have been the incidence density – rather than risk – of the event's occurrence in the domain at issue; and it would have been studied in terms of the causal incidence-density ratio, which determines the etiognostic probability, with allowance for this ratio's dependence on modifiers of its magnitude. (Chap. 6.)

This duality in causal objects of study implies its counterpart in the respective structures of the studies themselves. A prognostic study involves follow-up of a study cohort (a population closed to exit; once a member, always a member) enrolled as of prognostic T_0. An etiognostic study, in turn, is based on a case series from and a sample of the population-time constituting the study base formed by a dynamic (open-to-exit) population (with turnover of membership) over a span of time, the etiognostic T_0 characterizing the entire study base and hence both the case and base series. (Chaps. 6 and 7.)

The study by Etminan et alii, despite all those references to risk, actually was an etiognostic one. I'll address its particulars in Sect. 6.8, here merely noting that the report of the study would have called for major help from the peer reviewers and/or from the Journal's Editors. The particulars of this too I'll address in Sect. 6.8.

Significantly, there is only this one report of a pharmaco-etiognostic study in the 24 issues of *JAMA* constituting its January-June 2012 volume. This I deem to be out of proportion to the importance and commonality of medications' use as alternatives to illnesses as direct causes of patients' new complaints about sickness and of the commonality of medications' use also in the causation of illnesses. At the root of this discrepancy may be the still-common failure even to recognize etiognosis as a concern analogous to diagnosis and prognosis – as one of the three generic categories of a doctor's esoteric knowing about a client's health (Sects. 1.1, 3.1 and 3.2).

This semi-systematic, though only cursory, review of the recent contents of a very eminent 'medical' journal gives, as I've insinuated in the annotations along the way, a strong impetus for consideration and implementation of needed innovations in the journalism concerned with scientific medicine and medical science.

These needed innovations I address in Sects. 4.4 and 4.5, respectively, after having dealt with needed innovations of more proximal nature in Sects. 4.2 and 4.3; and after having, preparatory to all of these, addressed, in Sect. 4.1, the fundamental distinction that's to be made between scientific medicine and medical science.

3.4   Medical Education on the Needed Knowledge

As for the status quo of medical education, suffice it to take note of the outlines of this in my own faculty of medicine (at McGill University), focusing on the undergraduate medical education and the postgraduate programs in cardiology and adult respiratory medicine.

In the *undergraduate* education, the *objectives* are based (ref. 1 below) on:

two fundamental premises to ensure career-long excellence in whole-person care:

(1) The basic sciences and scientific methodology are fundamental pillars of medical knowledge and
(2) A physician fulfills two roles in service to the patient and society: that of a professional and of a leader, also termed Physicianship.

The objectives are organized by general competencies and principles deemed essential for Canadian physicians as elucidated by the Royal College of Physicians and Surgeons of Canada (CanMEDS 2005) and the College of Family Physicians of Canada (CanMEDS FM).

Reference 1: http://www.medicine.mcgill.ca/ugme/curriculum/objectives_en. htm, on 10/25/2012

The CanMEDS (ref. 2 below) actually specifies "A framework of essential competencies [and no principles] for Canadian specialist physicians" – which excludes practitioners of "family medicine" – with these elements:

Medical Expert/Clinical Decision Maker [e.g., ability to "demonstrate diagnostic and therapeutic skills [*sic*] for ethical and effective patient care"]
Communicator [e.g., ability to "listen effectively"]
Collaborator [e.g., ability to "consult effectively"]
Manager [e.g., ability to "utilize information technology"]
Health Advocate [e.g., ability to "identify important determinants of health"]
Scholar [e.g., ability to "critically appraise sources of medical information," and to "contribute to development of new knowledge"]

Reference 2: Frank J (2004) The CanMEDS project: the Royal College of Physicians and Surgeons moves medical education to the 21st century. In: Dinsdale HB, Hurteau G (eds) The evolution of specialty medicine 1979–2004. Royal College of Physicians and Surgeons of Canada, Ottawa, p 192

Those objectives of the undergraduate education – "to ensure career-long excellence in whole-person medicine" – are pursued in terms of a 4-year program. In it,

"Basis of medicine" is addressed from August of year 1 through November of year 2, starting with 4 weeks on "Molecules, cells & tissues" and ending with 14 weeks on "Pathobiology, treatment & prevention of disease." The rest of the education is labeled "Physicianship 1–4" and also "Physician apprenticeship 1–4." In it there is a 7-week period of "Introduction to internal medicine, pediatrics," later supplemented by 8 weeks of "Internal medicine" and also 8 weeks of "Pediatrics." And there is a 7-week "Introduction to surgery, radiology and ophthalmology," later supplemented by 4 weeks of "General surgery" and later yet by "Surgery sub-specialty." There is a 7-week "Introduction to family medicine, oncology, neurology and anesthesia," and later 4 weeks of "Family medicine rural" and "Family medicine urban." Etc. From this it is unapparent that "The objectives were organized by general competencies and principles deemed essential for Canadian physicians" (cf. above).

And upon graduation from all of this, the attained essential competencies (cf. above) are taken to provide the ability "to function responsibly, in a supervised clinical setting, at the level of an 'undifferentiated' physician."

The postgraduate program on *cardiology* "trains clinically mature and competent specialists with a broad level of expertise in the diagnosis and management of both routine and complex cardiovascular disease" (ref. 3 below).

Reference 3: http://www.medicine.mcgill.ca/postgrad/programs/cardiology.htm, on 10/25/2012

A prerequisite for entry into this program is 3 years of internal-medicine "training." And in the cardiology training that follows this, also of 3 years' duration, the accent indeed is on training, the acquisition of various "skills." Nothing in its description appears to harken back to the "fundamental premises" of the under-graduate education (above), nor to the "essential competencies for Canadian specialist physicians" (above).

Nevertheless, "Acquisition of an approach [*sic*] to performing *research*, and completion of a research project, are considered *essential* to a successful cardiology training period, and candidates will be expected to participate in a research project (basic, clinical or epidemiologic) in the area of cardiovascular disease over the three-year period" (italics added). (Cf. "essential competences for Canadian specialist physicians" in the capacity of "scholars," above.)

In *pulmonology* residency training, following 3 years in internal medicine, McGill has two major streams (ref. 4 below): Core Adult Respiratory Medicine and Respiratory Medicine Clinician/Scientist.

Reference 4: http://www.medicine.mcgill.ca/respdiv/print/12, on 10/25/2012

In the first one of those two streams the trainees spend 18 months in "clinical training," one feature of which is "extensive exposure to bronchoscopies and pleural procedures." As with the cardiology program, the description makes no reference to learning the gnostic knowledge-base of the discipline. "An additional 6 consecutive months are devoted to supervised clinical or basic science *research*" (italics added).

In the Respiratory Medicine Clinician/Scientist stream, the first 2 years are the same as in the Core Adult Respiratory Medicine stream. "In addition, there is a fully funded 3 year in research, designed to provide the trainees with further research training – in basic science (laboratory) or clinical-epidemiologic research methods."

I leave these descriptions without any annotations specific to them. But I'll outline what I see as the needed innovations in medical education and training in Sects. 4.1, 4.6, 4.7 and 9.1.

Part II
Whither From Here?

Chapter 4
Needed Innovations of the Knowledge Culture

Contents

4.0 Abstract

The knowledge culture of, and surrounding, medicine is in need of radical reformation. Fundamental to a sound knowledge culture of medicine would be a sound conception of *scientific medicine* – replacing the two mutually contradictory misconceptions of this that came to eminence in the twentieth century and still prevail – this together with appreciation of the generic nature of the requisite knowledge-base of medicine. Fundamentally needed, too, would be a rational taxonomy of the knowledge-dependent disciplines of medicine.

Textbooks of medicine should conform to those fundamentals; and in medical journalism, a distinction should be made between the genuine knowledge needs of practitioners and the concerns, ontal and epistemic, of researchers.

Undergraduate medical education should focus on that which constitutes the genuine 'medical common' across the disciplines of knowledge-dependent medicine, free of elements of skills-based medicine and medical science.

O. S. Miettinen, *Toward Scientific Medicine*, DOI 10.1007/978-3-319-01671-9_4,
© Springer International Publishing Switzerland 2014

In this Information Age, the knowledge-base of medicine – for dia-, etio-, and prognostic probability-settings – should be codified in practice-guiding gnostic expert systems.

4.1 Commitment to Genuinely Scientific Medicine

Medical science in the meaning of medical research serves to advance general – abstract; placeless and timeless – knowledge, for the advancement of the practice of medicine. It thus is application-oriented – 'applied' – science, not science for the sake of science; and it is quintessentially applied when aimed at the advancement, specifically, of the knowledge-base of the practice of medicine (Preface, Chap. 1). While in the service of medicine, medical science is not medicine, nor is medicine science; medical science is extrinsic to medicine – to physicians' practice of healthcare, that is.

While there scarcely is any notable confusion about this knowledge-pursuing essence of medical science (just as of any other science), in genuinely scholarly circles at least, appropriate conception of *scientific medicine* has remained remarkably elusive, even in medical academia. Pathognomonic about the confusion about this already are the surface facts that two very different conceptions of scientific (practice of) medicine came to eminence in the twentieth century; that both of them continue to prevail in medical academia; and especially, that both of them are characterized by absence of any conception of the role for *general medical knowledge* – that which quintessentially applied medical science is all about (cf. above) – in the practice of scientific medicine.

In 1910 was published the very influential '*Flexner report*' on education in medical schools (ref. 1 below). Its production was prompted by the American Medical Association, concerned to eliminate the large number of medical schools that it saw as producing quacks rather than real doctors. As Abraham Flexner – an educator but not a physician – reviewed medical schools in the U.S. and in Canada, he was inspired by the culture in them of research-based academia, introduced by the University of Berlin as of its very founding (by Alexander Humboldt), in 1810. And he was particularly inspired by this culture as he found it to prevail in the medical school of Johns Hopkins University.

Reference 1: Flexner A (1910) Medical education in United States and Canada. Bulletin Number 4. The Carnegie Foundation for the Advancement of Teaching, New York

The research in question was *laboratory* research, and Flexner emphasized what he took to be its role – critically important role – in medical education. The idea was that students' experiences in the research laboratories of medical schools provide for learning the scientific way of *thinking* (of the laboratory scientists). And the associated idea was that it is this scientific way of thinking that makes for successful problem-solving in the practice of medicine – of scientific medicine in this meaning of deploying, in the practice of medicine, the scientific way of thinking. Very notably, however, there is no trace of this doctrine in a very eminent contemporary book on doctors' thinking (ref. 2 below).

Reference 2: Groopman J (2007) How doctors think. Houghton Mittlin Company, Boston

By the end of the twentieth century this Flexnerian doctrine of 'thinking-based' scientific medicine had become rivaled by that of *'evidence-based'* medicine, in which the role of the patterns of thought of laboratory scientists is replaced by that of evidence from what I term meta-epidemiological clinical research. In expressly defined terms EBM is an ideology, or cult, spearheaded by David Sackett and his collaborators, one that makes the practitioner of medicine an actual scientist of sorts (ref. 3 in Preface): Each practitioner, on his/her own, reviews the evidence from original or review articles on an "answerable question" – "critically," according to set guidelines – and then applies this evidence – actually his/her own opinion about the meaning of this evidence – in decisions about patient care. The core idea (ref. 3 below) is anti-authoritarianism, replacement of deference to scientific experts by individual practitioners' evidence-influenced subjectivism, in the framework of what the movement's leaders define the practice of EBM to be, the first one of its five 'steps' being the asking of an "answerable" (rather than the first relevant) question.

Reference 3: Evidence-Based Medicine Working Group (1992) Evidence-based medicine. A new approach to teaching the practice of medicine. J Am Med Assoc 268:2420–2425

Without reference to this ideological movement, the adjective 'evidence-based' now has quite universal appeal, within medicine and beyond – much more than 'thought-based.' The appeal is intuitively so strong that there commonly is no felt need to explicate the idea, ultimately as to why this purportedly 'evidence-based' feature of someone's opinion or action makes it justifiable without any need to think about the nature of its foundation in evidence rather than learning or knowledge – and relevant values.

But science unarguably is about knowledge (cf. above), and therefore the most *fundamental innovation* now needed in the knowledge culture of medicine is general coming to grips, firmly, with the incontrovertible fact that commitment to genuinely scientific medicine – by the very nature of science, its quest for abstract knowledge (Gr. *epistēmē*) – is commitment to *knowledge-based medicine*, KBM, rather than thought-based medicine, TBM, or evidence-based medicine, EBM (see above). Remarkably, in the thick of today's commitments to actions and policies that are 'evidence-based,' even the term 'knowledge-based medicine' remains unheard-of.

In genuinely scientific medicine the knowledge that is deployed is, most broadly, of two very different kinds. Inviolable in science itself are the dictates of logic; and by the same token, scientific medicine is characterized by deployment of a known, *rational theoretical framework*. As an example of this, in genuinely scientific medicine the theoretical framework for diagnosis is not one involving likelihood ratios but, instead, probability functions (see Sect. 2.1). The terminus of strictly scientific medicine is the esoteric particularistic knowing – gnosis (dia-, etio-, pro-; Chap. 1) – it provides for, since more than this gnosis goes into rational decisions about actions in medicine, these added inputs – matters of valuation – being extrinsic to the domain of medical science.

Apart from the rational theoretical framework, an essential feature of genuinely scientific medicine naturally is that it has, in this framework, its *substantive knowledge-base from medical science* (for the needs of gnosis; Chaps. 1 and 2) – from quintessentially applied medical research. Fundamental to research for the knowledge-base of the practice of genuinely scientific medicine therefore is the rational theoretical framework of medicine (its practice); and it is fundamental to rational practice even in the absence of substantive knowledge from medical science. It is a framework defined as a matter of *general theory of medicine*, addressing general concepts and principles of medicine, and terminology besides.

A word about the respective roles of the scientist and the practitioner may be in order here. When a scientist works to advance the scientific knowledge-base of medicine, (s)he is working in the domain of medical science, as a scientist, not in the domain of medicine, as a doctor. And when a doctor is applying knowledge from medical science, (s)he is acting in the domain of medicine, as a doctor, not in the domain of medical science, as a scientist. Thus, as I noted above, application of scientific knowledge is not science; and thus, even scientific medicine is not medical science. Groopman (ref. 2 above), for example, writes that "Medicine is, as its core, an uncertain science" (p. 7); but the truth is that scientific medicine deploys, at its core, knowledge about the levels of uncertainty in its gnoses – gnostic probabilities derived from medical science.

In the present era of presumedly scientific medicine (either TBM or EBM; see above) – really pseudo-scientific medicine – the practitioner quite commonly has affectations or even the presumption of being a scientist, commonly wearing the scientist's laboratory coat in the practice. Related to this, the idea still officially is that involvement in research makes for a better practitioner (Sect. 3.4) and that "The basic sciences and scientific methodology are fundamental pillars of medical knowledge" (Sect. 3.4) – an echo of the Flexnerian doctrine. In the same vein, the

Editors of an eminent textbook of medicine open the book's first chapter with these words: "Medicine is a profession that incorporates science and the scientific method [sic] with the art of being a physician" (Sect. 3.1).

4.2   Redefinition of the Disciplines of Medicine

In its *The Evolution of Specialty Medicine*, The Royal College of Physicians and Surgeons of Canada (Sect. 3.4) lists the "specialties and subspecialties" it accredited in 2004. The Division of Medicine accredited 27 of these under Clinical Specialties, seven under Laboratory Specialties. The Division of Surgery accredited 12 specialties and subspecialties, without that clinical versus laboratory duality. In addition, the College accredited residency programs in 14 subspecialties for which it offered no examinations.

The division of entries between the Division of Medicine and the Division of Surgery seems to call for rearrangements. Shouldn't anesthesia be removed from Medicine and be under Surgery? And don't the purportedly surgical disciplines of obstetrics and gynecology, ophthalmology, otolaryngology, and urology have a medical knowledge-base, including about prognostic implications of the surgical interventions in them?

That conceptual duality of medicine-versus-surgery in evidence here and even in the very name of the College – implying that surgery is not medicine – actually is inconsistent with the realities of medicine. My *Dorland's* defines medicine as "the art and science [sic] of the diagnosis and treatment of disease and the maintenance of health," with no hint of surgical treatment being extrinsic to medicine. Nor is there this implication in the way medicine is defined in my *Stedman's*. In that quoted definition, "diagnosis and treatment of" should be replaced by "gnosis about and intervention on," where 'gnosis' refers to dia-, etio-, and prognosis, and 'intervention' refers to preventive and therapeutic ones without them being limited to non-surgical types (such as, e.g., radiotherapy).

In these terms, the central duality in the arts/disciplines of medicine is that constituted by the *knowledge-based* ones (concerned with gnosis, incl. prognosis about intervention effects) and *skills-based* ones (concerned with execution of interventions, e.g., radiotherapeutic and surgical ones). A discipline that is not concerned with the pursuit and attainment medical gnoses nor with execution of medical interventions is not a genuinely medical discipline – 'medical' microbiology, 'medical' histology/cytology, and 'medical' chemistry, for example. These laboratory disciplines are paramedical in the meaning of being subservient to medicine; they produce data for entry into gnostic profiles, not gnoses proper. Diagnostic radiology should be viewed as one of these laboratory disciplines, as providing data inputs (readings from pictures) to diagnostic profiles rather than producing diagnoses proper.

In the taxonomy of knowledge-based, gnosis-oriented medical disciplines the distinctions should reflect major differentiations of the requisite knowledge-base. Central in gnosis naturally is diagnosis; and the requisite knowledge-base of diagnosis generally is highly differentiated according to what sickness is the prompting for the diagnostic pursuit. An example of the needed, rationally construed *diagnostic disciplines* would be that concerned with diagnosis in the context of an adult with acute chest pain and/or dyspnea, very different in its requisite knowledge-base from that concerned with diagnosis on the prompting of complaint of low-back pain, for example. So there should be separate disciplines for these two presentations for diagnosis, among many others.

And while these two would be examples of disciplines concerned with diagnosis, very different disciplines, as for the requisite knowledge-base, should be seen to be needed for prognosis, notably intervention-prognosis, in the various rule-in diagnosed thoracic and lumbar illnesses. As just one example of this, a doctor competent in the pursuit of differential diagnosis in the context of acute chest pain and/or dyspnea – among impending myocardial infarction ('unstable angina'), myocardial infarction, pericarditis, pulmonary embolism, pneumothorax, etc. – is not, ipso facto, competent in formulating prognoses (intervention-conditional) in rule-in diagnosed cases of any of the illnesses in question. To this end needed are *prognostic disciplines*, the one concerned with cardiac illnesses very different in its requisite knowledge base from that concerned with pulmonary illnesses.

All of this is to suggest that the Royal College should replace its existing two devisions by Division of Gnosis and Division of Intervention. From the vantage of concern for scientific medicine, the focus is on the former, with the Canadian disciplines of it in 2004 constituted by the then 27 "clinical" disciplines addressed by the division of "medicine." Only one of these is defined by a genre of sickness (as a diagnostic challenge) – psychiatry. As for the others, the set leaves quite unapparent the principles of inclusion/exclusion. Particularly notable is the inclusion of community medicine, without subdisciplines, among clinical disciplines of medicine (cf. Preface).

4.3   Innovation of Textbooks of Medicine

Along with the needed major innovation of the taxonomy of the disciplines of knowledge-based medicine (Sect. 4.2 above) there naturally is the need for the corresponding, major innovation of the textbooks for this segment of medicine.

Given the spectrum of rationally-defined disciplines of knowledge-based medicine, this in the context of commitment to rational, ultimately scientific medicine (Sect. 4.1), needed for a start is introduction to all of this, an introductory textbook: *Introduction to Medicine*. It might be bi-partite: Part 1 would address, in a rigorously scholarly way, the general concepts and principles of medicine, together with the appropriate terminology for the concepts; and Part 2 would be devoted to delineation of the spectrum (rational) of differentiated disciplines of knowledge-based medicine, including the discipline-specific distributions of types of client presentation and physician action.

The concepts and principles addressed in this book would need to encompass, for one, the logic-dictated ones of the general theory of medicine, including decision theory (of statistics and economics) as to its inapplicability to medicine. These would need to be supplemented by principles of medical ethics and medical professionalism. Under medical professionalism, suitable attention would need to be given also to the functional interrelations among the differentiated disciplines, including those among primary, secondary, and tertiary care; those of diagnosis and prognosis (Sect. 4.2); those of prognosis and intervention (Sect. 4.2); etc.

This book, notably if Part 2 indeed would be included, might initially need to have several, say half-a-dozen, authors; but, by the shared relevance of its contents across the various disciplines of knowledge-based medicine, later editions could well have only a single author – akin to Osler's book on medicine (Sect. 3.1), which came out before the differentiation of medicine into its modern multitude of component disciplines. For, once the medical common has been well defined, all of it will have to be fully mastered by practically all professionals of medicine – most especially by all authors of any textbook on it.

Beyond that broadly introductory yet suitably focused textbook on the genuine medical common across the disciplines of knowledge-based medicine – in its contents radically different from the undergraduate curricula of today's medical schools (Sect. 3.4) – most urgently needed would be a radically novel type of textbook on *Primary-care Medicine* – as a replacement of what still are undifferentiated and hence ever-enlarging textbooks of medicine-not-otherwise-specified (à la *Davidson's* and *Cecil*, i.a.; Sect. 3.1), very wanting in their representation of the knowledge-base of primary-care medicine and empty of content on the skills (quite limited) needed in this discipline of medicine.

In this book on primary-care medicine the very first part would delineate a (practically) comprehensive set of the reasons – mainly gnostic but also interventive – why people seek care from a primary-care doctor. It would also characterize these client presentations according to whether primary care actually is called for, initially at least; and if not, according to the discipline or care setting to which the case is to be referred, forthwith. Delineation of the nature of sicknesses would be eminent in this part.

Another broadly orientational part would then give succinct definitions for all of the illnesses that can be causal to the sicknesses of concern for care in this front-line discipline of medicine, supplementing this by descriptions of the skill-requiring procedures – diagnostic and interventive – that reasonably can be involved in primary-care medicine.

The next part would address primary-care diagnoses. It would focus, one by one, on (practically) all of the types of patient presentation – as to chief complaint together with the patient's demographic category (broadly) – that call for primary-care diagnosis, including as the basis for referral to other disciplines, presentations such as 'Adult with acute substernal pain,' or 'Child with cough and fever.' Under each of these patient-presentation rubrics, the book would specify all of the information on the case that is relevant to ascertain from history and physical examination (Sect. 1.1), perhaps in the form of a clinical-diagnostic questionnaire to be filled out; and it would specify the complete differential-diagnostic set (inclusive of non-illness causes of the sickness; Sect. 1.1).

While this already would be very important for the primary-care doctor to know, the critical element of needed knowledge is not yet there: the set of profile-conditional probabilities for the presence of each of the illness members of the differential-diagnostic set, conditional on the premise that the sickness is a manifestation of illness (rather than a non-illness circumstance). For the possibilities of most urgent care, if any, and for otherwise important possibilities, the textbook (electronic perhaps) would give the respective diagnostic probability functions, specific to the presentation domain (Sect. 2.1). If those functions are not yet available from diagnostic research (Sects. 5.7 and 5.8) they need to be derived for the book from the domain-specific diagnostic experts' tacit knowledge (Sects. 4.8 and 9.1). If the function-implied probabilities are not extreme enough for action (diagnosis-conditional treatment or referral for further diagnostics), the doctor needs to learn from the book what laboratory tests could serve to provide for sufficiently conclusive diagnoses; and (s)he needs to get the corresponding post-test diagnostic function(s) from the book. This needs to be supplemented by knowledge relevant to etiognosis about the non-illness elements in the differential-diagnostic set.

The remaining part would address prognoses in primary care, both treatment-effects prognoses and treatment-conditional prognoses in reference to practically rule-in diagnosed cases of particular illnesses, supplementing this by its counterpart for primary-care preventive medicine. The respective organizations of the texts would be by type of illness diagnosed and type of illness that is of preventive concern.

Today's textbooks of medicine-not-otherwise-specified thus are (Sect. 3.1), in their form as well as their contents, a far cry from what really is needed for primary-care medicine – as a textbook "Designed for the use of Practitioners and students of Medicine" (cf. Osler in Sect. 3.1).

Textbooks for secondary- and tertiary-care knowledge-based medicine require equally radical reconfigurations. They each need to be specific to a particular discipline or setting of care.

4.4 Innovation of Journalism on Medicine

Suitably focused journalism on medicine – rational medicine evolving into genuinely scientific medicine – naturally would be directed to *practitioners* of medicine, and it thus would deal with medical science only in terms of disseminating scientific advances in the knowledge-base of practice, and this, even, only insofar as the advance is in the knowledge-base of the discipline(s) of medicine to which the journal is directed. Such content of science would be understood to belong in general medical journals (such as, e.g., *JAMA*) only in the highly exceptional instances in which the new knowledge is of concern in (practically) all disciplines of medicine (Sect. 3.3).

In this newly enlightened culture of expressly medical journalism, not even 'conclusions' from studies – even review-type, derivative studies – would be presented to practitioners (as though constituting new knowledge), much less the evidence behind these. Thus, practitioners' exposure to, for example, the 'statistical methods' (incl. charades such as 'sample size determination') of these studies would no longer be there to unwittingly promote among them the illusion that they too are scientists, supposed to be qualified to critically evaluate the evidence behind those 'conclusions' (cf. EBM in Sect. 4.1), which commonly are not even of the form of genuine conclusions (Sect. 3.3).

For this yet-to-be-introduced remedied culture of journalism directed to practitioners of medicine, a paradigm would be that for practitioners of the arts of agriculture – which now are much more scientific than those of medicine. This journalism helps farmers internalize and put to practice new scientific knowledge, without them having to follow and critically evaluate the scientific evidence and conclusions reported by fellow farmers or by dedicated agricultural scientists, and then having to develop their own, subjective opinions about the practice implications of the evidence (cf. EBM in Sect. 4.1).

4.5 Innovation of Journalism on Medical Science

The place of scientific evidence in the suitably reformed culture of journalism surrounding medicine would be in journals that expressly are ones of medical science, directed not to practitioners of medicine but solely to medical scientists;

and the culture of these journals in respect to evidence would be quite different from what it now is in 'medical' journals. The culture of these journals would be in tune with the genuine culture of science; it would not be dictated by editorial policies at variance with this.

Gone would be the culture in which the journals' editorial policies are quite intolerant of critical discussion of any evidence they have published; in which allowed is only a short time for the submission of critical commentaries with severely restricted number of words in these; and in which the journals only selectively publish these, never expecting the authors to modify their 'conclusions.'

In line with the genuine culture of science, the reformed journals of medical science would actively cultivate discussion of the evidence they've published, not only fully tolerating it but also encouraging and inviting it. In this spirit, they would set no arbitrary limits to the size, or time for the receipt, of commentaries on evidence they've published, nor would they exercise selectivity in the publication of the commentaries they receive. For in the inherently egalitarian culture of genuine science, *no authority* regulates the flow of ideas among scientists, especially not authority rooted in one's position in the journalism rather than exceptional competence in and impartial dedication to advancement of the science in question.

Once this culture of journalism in medical science would be in place, new evidence would be accepted for publication on the merits of the objects and methodology of the study, with no role for the result in this; for, this policy is needed to forestall publication bias entering into the published evidence. This means irrevocable acceptance for publication on the basis of the study design (as for both objects and methods), before any implementation of this design. But this does not mean publication of the study report as it first is presented, for it remains for the editors to assure freedom from such errors of reporting as now remain commonplace (Sect. 3.3).

In this suitably reformed culture of medical-scientific journalism there would be a fundamental duality in the types of scientific journal, one type dealing with nothing but a *presentation, interpretation, and evaluation of evidence*, the other type focusing on the even more challenging topic of *forging knowledge from the evidence*, a process in which free public discourse by members of the relevant scientific communities is even more essential than in coming to grips with the nature and merits of the available evidence. In fact, with the evidence and its critical discussion the background for this inferential mission, the requisite process is entirely a matter of public discourse among the topic-specific scientific experts. This latter type of journalism in medical science deserves to be separate from that focusing on evidence as it is so very different, so much so that it still remains essentially absent from quintessentially applied medical science, even though the ultimate purpose of the research is not the attainment of evidence but of knowledge (Sect. 4.1).

As the simplest and most obviously needed major innovation to be mentioned here, journalism of medical science should be divorced from medical commercialism. The scientific journals should not continue to be medications-advertising and subscriptions-hogging businesses; to be impartial scientifically, they ought to be *publicly financed*.

4.6 Innovation of Education for Medicine

As the differentiation in medicine into its component disciplines developed in the twentieth century (and continues in this 21st, still unguided by critical reasoning; Sect. 4.2), the corresponding needed innovation would have been discontinuation of the education of "undifferentiated physicians" (Sect. 3.4) – 'medical doctors,' MDs. But this has not happened, and much of what is wrong in today's medicine is sustained by the perpetuation of this MD concept.

Logic – this inviolable disciplinarian of thought in science and genuinely scientific medicine (Sect. 4.1) – allows a disinterested observer to readily understand that, should all of the educational content behind the MD degree be relevant to practice in whichever discipline of medicine, then all competent practitioners of medicine, regardless of discipline, would continue to master all of it or, where needed, an updated version of it – and all of them could teach all of it, even as a faculty of one. The prevailing reality is, however, that practitioners of medicine, however competent, generally retain practically no concrete idea of what they once knew – the chemical meaning of 'fat' or 'salicylic acid,' for example. They would fail almost all, if not all, of the examinations they once passed, to say nothing about the contemporary counterparts of these. The vast bulk of the content in today's medical-school curriculum (Sect. 3.4) is irrelevant for a career in whichever discipline of modern medicine.

While the first needed innovation of medical education would be abolition of that MD concept, a related needed other innovation would be the introduction of the succedaneous BMC – Bachelor of Medical Common – degree and education. This would be wholly science-free and focused on that which truly is of common concern across all of the disciplines of knowledge-based medicine – the contents of Chaps. 1 and 2 above, for example, but very little of today's MD education and of Part III of this book, for example. This BMC education would be relevant also for the only discipline of medicine that now can be prepared for without first acquiring the MD degree – dentistry, that is. The core content of this education would be that of the *Introduction to Medicine* textbook sketched in Sect. 4.3.

This education in the genuine medical common – quite short – would be immediately followed by education in the student's chosen discipline ('specialty') of knowledge-based medicine – primary-care medicine, for example, the elements of which also are sketched in Sect. 4.3. And no longer would 3-year 'training' in general 'internal' medicine be a necessary preparation for its subdisciplines (cf. Sect. 3.4), just as corresponding investment in 'external' medicine is not necessary preparation for a career in dermatology, for example.

This pre- and post-graduate education for knowledge-based medicine would not call for any preparatory, *'pre-med' studies* – of biology, for example. Removal of this requirement would be a needed innovation in countries such as the U.S. and Canada.

These needed innovations in the culture of medical education may not be introduced by the faculties of medical schools of the prevailing type, or by their practicing graduates, not even in response to whatever futuristic teachings directed to these, such as this book. For, as Upton Sinclair noted, "It is difficult to get a man to understand something, when his salary depends on not understanding it." Needed may be societal imposition.

4.7 Innovation of Education for Medical Science

Education for medical science is in this book of concern specifically for research to advance the *knowledge-base* (scientific) of the practice of rational and increasingly scientific medicine, exclusive of 'basic' medical research.

Just as the post-BMC education (not 'training') for the practice of knowledge-based medicine – scientific at least in the meaning that its theoretical framework is rational (Sect. 4.1) – would have to do with content relevant to a particular discipline of medicine (Sect. 4.6 above), so the post-BMC preparation for quintessentially applied medical research would need to be directed to research to advance the knowledge-base of a particular knowledge-based discipline of medicine (diagnostic or prognostic; Sect. 4.2). For, the graduates of such education would typically be academics who not only conduct and follow research but also provide education for future academics in a particular discipline of medicine. They need comprehensive, intimate knowledge about the need for and state of practice-relevant knowledge in their particular discipline of medicine, in teaching of and priority-setting for the research, first and foremost.

This investigator-education would not be an alternative to that leading to practice of medicine in one of the knowledge-based disciplines of medicine; it would be a major (sic) addition to it. And just as the proper education for practice would be bi-phasic in the sense that the undifferentiated medical common is followed by that which is discipline-specific, so the proper research education following the practice-oriented education also would be bi-phasic, by the same principle. The first phase would be the interdisciplinary *gnostic-research common* – to which this book is an introduction of sorts – while the rest of the education would, again, be discipline-specific.

The students of clinical research in this yet-to-be-introduced cultural context would not only have the background preparation in the medical common and in the subject-matter of their particular discipline of medicine; they would also have the additional background of having studied the theory of epidemiological research (Preface). Besides, the research (gnostic) would be understood to require, even more than does epidemiological research, probability theory and statistics as areas of preparatory study – as 'pre-med' subjects of study. 'Pre-med' studies in biology, for example, would be deemed to be irrelevant, just as for the practice of scientific medicine (Sect. 4.6 above).

In this novel culture, providing for the advent of genuinely scientific medicine, competence in quintessentially applied medical research would be cultivated with the necessary seriousness, and not treated as something that practitioners of medicine also understand. There would be no place for such flippant ideas about medical science as were presented in the section on prevailing medical education, Sect. 3.4. No longer would it be held that "The basic sciences and scientific methodology are the fundamental pillars of medical knowledge." No longer would the education of practitioners involve token pursuits of competence as a "medical scholar" and, to this end, process elements such as "Acquisition of an approach [sic] to performing research, and completion of a research project," or "6 consecutive months ... devoted to supervised clinical or basic [sic] science research" (Sect. 3.4). These schemes to cultivate the idea that a practitioner of modern medicine is a medical scholar, meaningfully educated in medical science, would be left behind as intellectual aberrations of medicine's pseudo-scientific past.

4.8 Development of Expert Systems for Medicine

When accepted truly becomes be the fundamental idea that most of the disciplines of medicine are to be viewed as principally knowledge-based (rather than thought-based or evidence-based, or skills-based) professions of healthcare (Sects. 4.1 and 4.2), in which doctors bring general medical knowledge to bear on ascertained facts on their clients to gain esoteric knowing about their health (Chap. 1), and when appreciated also becomes that the requisite knowledge-base – even within the constituent disciplines of knowledge-based medicine – is very complex (Chaps. 1 and 2), a challenging question of needed innovation – at the very core of medicine – arises.

The question is this: How is the existing state of the discipline-specific knowledge-base of medicine to be codified and made deployable in the practice of the discipline in question? This question arises because the knowledge presented in existing textbooks is woefully inadequate (Sects. 3.1, 3.2 and 4.3); because it isn't codified anywhere else either; and because it wouldn't be subject to being learned and memorized by practitioners of medicine even if it were to be codified (cf. Chap. 2).

The orientational answer to that question is quite obvious, recognized from the dawn of this Information Age, decades ago: practice of medicine (like commercial aviation, i.a.) should be guided by computer-based *expert systems*. The initial focus in this development has been, naturally enough, on diagnostic expert systems, but without actual success (ref. 1 below).

Reference 1: Wolfram DA (1995) An appraisal of INTERNIST-I. Artif Intell Med 93–116

Success in the development of diagnostic, and also etiognostic and prognostic, expert systems is predicated on the understanding that the necessary form of the knowledge-base of clinical medicine is one of gnostic probability functions, GPFs (Chap. 2). And it also requires understanding of how gnostic experts' tacit knowledge – dia-, etio-, or prognostic – can be garnered in the form of GPFs, before as well as after the appropriate gnostic research. For this there now is the requisite know-how, outlined in Sects. 9.1 and 9.2.

So the needed innovation now is to get on with the development of the GPFs for the various disciplines of knowledge-based medicine. Once the requisite GPFs are available for a given one of these disciplines, through a user-friendly expert system, all practitioners in this discipline can practice, as far as the critically important matter of the deployment of available relevant knowledge is concerned, on the level of quality of the typical expert in the discipline of medicine at issue.

There is a learned book on how medicine now, in this Informative Age, could be fundamentally improved (ref. 2 below). It, however, does not embody the vision put forward here, which in essence is this: In this Information Age, a consumer of healthcare should not have to rely on the knowledge personally possessed – held in neurospace – by his/her doctor; (s)he should be able to draw, through his/her doctor from cyberspace, the needed knowledge – the best available knowledge in the doctor's particular discipline of medicine, with no need for a 'second opinion.' And this is not here presented as a dream or a prediction but as a vision for a needed, immediately implementable, very major innovation of the knowledge culture and practice of medicine – practice that is excellent not only medically but, secondary to this, economically as well. More on this in Chap. 9.

Reference 2: Topol E (2012) The creative destruction of medicine. How the digital revolution will create better health care. Basic Books, New York

Part III
Medical Science for Scientific Medicine

Chapter 5
Original Research for Scientific Diagnosis

Contents

5.0 Abstract

Even though the pursuit of diagnosis is the pivotal element in the practice of medicine, it remains quite poorly understood. Even the general concepts of and related to diagnosis still commonly are quite malformed, to say nothing about the general principles of diagnosis. Consequent to this, the requisite research to advance the knowledge-base of scientific diagnosis remains seriously misguided.

In research for genuinely scientific diagnosis, the objects of study are imbedded in two generic types of function. The principal one of these is a *diagnostic probability function*, expressing the probability of the presence of a particular illness as a joint function of a set of diagnostic indicators in a particular domain for the diagnostic challenge (Sect. 1.2). The other type is a *test performance function*, expressing the probability that a particular type of testing will produce a result providing for practical rule-in or rule-out diagnosis, this as a joint function of the pre-test diagnostic indicators.

The study-specific form of each of these two types of function is the result of the study's *objects design*, which dominates the study's *methods design*.

O. S. Miettinen, *Toward Scientific Medicine*, DOI 10.1007/978-3-319-01671-9_5,
© Springer International Publishing Switzerland 2014

5.1 Diagnosis-Related Concepts à la *Dorland's*

A student preparing for excellence in medical research of the meta-epidemiological clinical kind (Preface) needs to have the disposition of a *medical scholar*. The most proximal feature of this is scholarly inquiry into the generic concepts and terms of medicine – critical examination of the prevailing concepts and the corresponding terms, and judicious adoption of ones that would best serve as the elements in his/her critical thinking concerning the research, critical study and adoption of the principles of this research.

In respect to diagnostic research, the very first concern naturally is the essence of diagnosis – the concept of diagnosis, specified by its definition – together with the concepts needed in the definition of the concept of diagnosis.

In one of my medical dictionaries – *Dorland's Illustrated Medical Dictionary*, 28th edition, 1994 – the denotation of '*diagnosis*' is presented as:

1. The determination of the nature of a case of a disease.
2. The art of distinguishing one disease from another.

This definition follows the etymologic point that the Greek word *gnōsis* means 'knowledge,' and associated with this definition of diagnosis per se is the definition of *differential diagnosis,* as:

> the determination of which one of two or more diseases or conditions a patient is suffering from, by systematically comparing and contrasting their clinical findings.

Among the constituent diagnosis-related concepts in these definitions, the salient one is that of *disease*. The dictionary defines it as:

> any deviation from or disruption of the normal structure or function of any part, organ, or system (or combination thereof) of the body that is manifested by a characteristic set of symptoms and signs and whose etiology, pathology, and prognosis may be known or unknown.

"Condition" as an alternative to disease (in that definition of differential diagnosis) the dictionary does not define, giving only "to train; to subject to conditioning."

Clinical (in that definition of differential diagnosis) the dictionary defines this way:

> pertaining to a clinic or to the bedside; pertaining to or founded on actual observation and treatment of patients, as distinguished from theoretical or basic sciences.

Having read all of this, the scholarly student pauses as a matter of course, to think critically about it. For (s)he understands that at issue here are concepts as central and pivotal as any to medicine and, hence, to all of the research to advance the knowledge-base of this core of medicine; and (s)he also understands that, as a scholar, (s)he is not to go by mere 'received truths,' not even in respect to the definitions of the most central concepts of medicine.

Accordingly, (s)he is more amused than moved by what the Chief Lexicographer of that 28th edition of *Dorland's*, D. M. Anderson, says in its Preface. The opening words are these:

> During what is now nearly a century of existence, *Dorland's Illustrated Medical Dictionary* has been the outstanding authoritative guide to the language and usage of medicine and related fields.

And a bit later there is this:

> it has been our purpose to provide you, the user, with an authoritative and current guide to the vocabulary of medicine; we hope that you will agree that this dictionary realizes that purpose.

The scholarly student has a strong aversion to this attitude of authoritarianism in matters scholarly, and (s)he rejects off-hand the notion that a medical dictionary is to be thought of as a guide primarily to the "vocabulary," rather than the concepts, of medicine. To wit, any prospective researcher concerned to help advance the knowledge-base of medicine has such terms as 'diagnosis' and 'disease' and 'clinical' in his/her vocabulary well before ever consulting a medical dictionary, but the corresponding tenable concepts (s)he may not have even after having consulted a purportedly "outstanding authoritative" medical dictionary about these.

"Read not to contradict, nor to believe, but to weigh and consider." So Francis Bacon counseled us all (ref. below). Having weighed and considered those definitions by *Dorland's* above, the scholarly student does incline to contradict, I presume, even when this wasn't the purpose of the readings.

Reference: Bacon F (1999) The essays or counsels civil and moral. Edited with an Introduction and Notes by B. Vickers. Oxford University Press, Oxford

Here's how I, after much weighing and considering, think about the diagnosis-related concepts behind the terms in those definitions.

Starting from the dictionary's definition of *disease*, I take a suitably edited version of it to be simply this, without change of meaning:

> any somatic anomaly manifest in a characteristic type of sickness.

For, "any somatic anomaly" is the meaning of the first 23 words in that dictionary definition; and, rationally, it is not definitional of any entity that something about it "may be known or unknown." Somatic anomaly does not subsume something that, while perhaps unusual, nevertheless is normal in the circumstances, as is the physiological condition attending to, say, early pregnancy (manifest in 'morning sickness'), intense physical exertion (manifest in 'athlete's sickness'), or low level of atmospheric concentration of oxygen (manifest in 'altitude sickness'). But: manifestation in a characteristic set of symptoms and signs is by no means true of diseases in general. If it were, latent disease – a latent case of a cancer, say – would be a contradiction in terms. Moreover, if overt cases of each particular disease would manifest, by definition, "a characteristic set of symptoms and signs," diagnosis would present no challenge at all; it would reduce to mere pattern recognition of syndromes of disease-specific sets of overt manifestations, which most definitely is not the case.

With these understandings that dictionary definition of disease gets to have this edited and corrected form (with seven words, instead of 45):

any somatic anomaly potentially manifest in sickness.

Pregnancy is not a somatic anomaly even though it is potentially manifest in sickness, and situs inversus is a somatic anomaly but it is not potentially manifest in sickness; neither one satisfies this definition of disease. By contrast, a latent case of cancer not only is a somatic anomaly but it also has the potential of progressing into an overt, clinically manifest case of the cancer, the manifestation in these terms being sickness (in contrast to the merely subclinical, radiographic manifestation of situs inversus); it thus is a case of disease.

Likely in agreement with these editings and intended-to-be corrections of what the "authoritative" dictionary of medicine gives as the denotation of 'disease,' the budding clinical scholar might, and indeed should, feel that one problem about that definition still remains unresolved. The question is whether that which the definition above gives as the essence of disease is unique to disease, or whether it is shared by other types of ill-health that also are of concern in diagnosis? For, a scholar expects a definition to specify that which is true of each instance of the thing being defined and unique to it.

Trisomy 21 is a somatic anomaly regularly manifest in Down's syndrome; and the sickle cell trait (sicklemia) is a somatic anomaly commonly manifest in 'aplastic crisis,' for example. But trisomy 21 and the sickle cell trait are congenital *defects*, not diseases; and examples of acquired defects with potential (incl. regular) manifestation in sickness include cirrhosis of the liver and status post hysterectomy, for example.

Furthermore, fresh subluxations of lumbar vertebrae and fresh fractures of bones are somatic anomalies with potential for manifestation in sickness. They, however, are examples of neither diseases nor defects but of *injuries*.

Budding clinical scholars, I think, might well – upon the needed weighing and considering – come to share my view that diagnosis in medicine is about ill-health of whatever kind, about *illness*, with 'illness' the appropriate antonym of 'health.' And further, that the principal subtypes of illness are disease (L. *morbus*), defect (L. *vitium*), and injury (L., Gr. *trauma*) – disease being a process-type anomaly ('disease process'), defect being an anomalous state (stable, more or less), and injury being a process that differs from disease in its etiology/etiogenesis. (The outcome of the course of an injury, when not fatal, commonly is a sequela, inherently a defect, and this type of outcome is common in some diseases as well.)

In these terms, the seven-word definition above actually is not that of disease, specifically, but of *illness* in general. And it thus is of interest to look up the definition of illness in my *Dorland's*:

> a condition marked by pronounced deviation from the normal healthy state; sickness.

Suitably edited, without change of meaning, this definition is:

> any pronounced ill-health; sickness.

But from the scholarly standpoint, if a condition of ill-health that represents "pronounced" (whatever may be the meaning of this) ill-health is an illness, then it is, specifically, pronounced illness, distinct from less-than-pronounced illness, without pronounced ill-health alone being an illness.

Dorland's brings up *sickness* in its definition of illness (above), though not when defining disease. For this its definition is:

> any condition or episode marked by pronounced deviation from the normal healthy state; illness.

This is followed by definition of "aerial s." all the way to that of "x-ray s.," 39 sicknesses in all. But: "morning sickness," for example, is not a "pronounced deviation from the normal healthy state." And: "sleeping sickness" this dictionary defines as "increasing drowsiness and lethargy, caused by ..."; it is the overt manifestation of the diseases causal to it. All of the 39 named sicknesses in *Dorland's* are defined this way – by their overt features together with a particular cause of these, whether a somatic anomaly or the circumstances. But sickness – or whatever phenomenon – is what it is regardless of its causation, known or unknown. Thus, for example, nausea-cum-vomiting is a sickness regardless of whether it manifests an illness or intolerance of motion or whatever else. Sickness is *overt unwellness*, whatever the causation of it.

Finally, as for the terms involved in or related to the definition of diagnosis in *Dorland's*, someone preparing for excellence in *clinical* research takes a critical notice of the specification of the denotation of this adjective as it is given in that dictionary (quotation above). Even though the word's etymology has to do with bed (Gr. *klinikos*, 'bed'), this has little to do with its various contextual meanings in medicine. Clinical medicine deals with individuals, one at a time, distinct from the population focus in community medicine. Clinical findings are findings from history and physical examination, distinct from laboratory findings. Clinical research is research for the advancement of clinical medicine, 'basic' research with this aim included, distinct from epidemiological research (incl. 'basic' research for community medicine).

5.2   Concepts of Diagnosis Proper à la *Dorland's*

Now that we've worked on the terms and concepts that are needed for tenable conception and definition first of diagnosis per se and then differential diagnosis, we're ready to return to the denotations of these two terms as they are presented in my *Dorland's* (Sect. 5.1 above), engaging in their critical examination.

Even though the dictionary specifies the Greek word for 'knowledge' as being etymologic to 'diagnosis,' neither the meaning it gives to 'diagnosis' nor that which it gives to 'differential diagnosis' is about knowledge. Instead, diagnosis purportedly is a "determination" and an "art," while differential diagnosis is said to be a "determination" but not an "art."

While the determination in one of the purported two meanings of 'diagnosis' is said to concern "the nature of a case of disease," and the art in the other one is said to deal with "distinguishing one disease from another," it appears that they really are meant to express the same thing. For, presumably, determination of the nature of a case of 'disease' is tantamount to distinguishing one possibility from another.

Something different – additional – is said about the determination under "differential diagnosis" ("the determination of which one of two or more diseases or conditions a patient is suffering from.") Added is a specification of the means to this end: "by systematically comparing and contrasting their clinical findings." Also added is the specification that the context of differential diagnosis is a patient's suffering from something.

Whether this specification of process is meant to distinguish differential diagnosis from diagnosis-not-otherwise-specified is left unstated. But given that this "systematically comparing and contrasting" is taken to be the process of differential diagnosis, it presumably is "the art of distinguishing one disease from another" as a matter also of diagnosis-not-otherwise-specified. And it remains quite unclear whether really meant is that a difference between differential diagnosis and diagnosis n. o. s. is that only in the former context is the client suffering from some disease.

The authors of an "authoritative" dictionary of medicine evidently are seriously confused not only about such central objects of medicine as disease, illness, and sickness (Sect. 5.1 above) but also about the most central concept of medicine itself, namely diagnosis. This is highly instructive about the level of scholarship that now generally underpins the MD degree, about how desperately needed are major innovations in the knowledge culture of medicine for genuinely scientific medicine to come about (Chap. 4).

It wasn't a matter of deep medical scholarship but, instead, an obvious truism, taken as such in Chap. 1 already, that a doctor caring for a client needs to achieve esoteric *knowing* about the client's health, by bringing general (abstract; placeless and timeless) medical knowledge to bear on particularistic facts on the client. This special kind of knowing – gnosis – is first and foremost *diagnosis*: a physician's esoteric knowing about the current presence/absence of a particular illness in a client. The particularistic facts about the client are the elements in the *diagnostic profile* of the case; and the needed general medical knowledge is about the *probability* that the illness at issue is present in any given instance of this profile, about the proportion of instances of the profile in general such that the illness at issue is present (Sect. 1.2).

When, as is usual, the client actually is a patient (Sect. 1.1), the complaint (about sickness) implies the corresponding set of possible underlying illnesses or other causes of it, the *differential-diagnostic set* (Sect. 1.2). But this does not call for some special art of "differential diagnosis" (cf. above), only diagnosis about each of the illnesses in the set as possible causes of the sickness. This contrasts with a-priori focus on a particular illness in the absence of any prompting complaint, as when addressing possible latent presence of a particular illness (Sect. 1.2).

When the client presentation is one of sickness (other than a syndrome) and thus implies the need to pursue diagnosis about each of several illnesses, it is not that a rational diagnostician engages in "systematically comparing and contrasting their clinical findings." At hand is a single set of findings, constituting the diagnostic profile of the case, and this is not to be compared and contrasted with anything, nor are the manifestations of the different illnesses compared (as to how well they match with the clinical profile at hand). Rather, the diagnostic question in respect to each of the possible illnesses at the root of the sickness is about the probability of its presence, given the diagnostic profile of the case, the diagnosis about it being awareness of – knowing about – the probability of its presence (Sect. 1.2).

5.3   Others' Misapprehensions about Diagnosis

I must not leave the impression that *Dorland's* is an exceptionally unscholarly dictionary of medicine. I therefore need to touch upon my other dictionary in this genre, *Stedman's Medical Dictionary Illustrated in Color*, the 28th edition (1995) of this, characterized in its Preface as an "authority on medical language."

For '*diagnosis*' the meaning is given as: "The determination of the nature of a disease. SYN diacrisis." Attached is the note that Greek *diagnōsis* means "a deciding." But for *differential diagnosis* there is no definition. Said only is that 'differential' means "Relating to, or characterized by, a difference; distinguishing."

And as for '*disease*,' which purportedly always is at issue in diagnosis, this dictionary defines the meaning of the word as:

1. An interruption, cessation, or disorder of bodily functions, systems, or organs. SYN illness, morbus, sickness.
2. A morbid entity characterized usually by at least two of these criteria: recognized etiologic agent(s), identifiable group of signs and symptoms, or consistent anatomical alterations. SEE ALSO SYNDROME.
3. Literally, dis-ease, the opposite of ease, when something is wrong with the bodily function.

'*Disorder*' is said to mean "A disturbance of function, structure, or both, resulting from genetic or embryologic failure in development or from exogenous factors such as poison, trauma, or disease." The meaning of '*morbid*' is given as "Diseased or pathologic," with the note that the Latin word *morbus* means "disease." And for *syndrome* the meaning is given as: "The aggregate of signs and symptoms associated with any morbid process, and constituting together the picture of the disease." Picture is left undefined.

In these statements, again, absence of scholarly, critical thinking about medical concepts – most elementary ones of these – is manifest in a number of ways:

1. 'Determination' in that statement purporting to give the meaning of 'diagnosis' is a noun that derives from the verb 'determine,' which according to my OED means "find out or establish precisely." But surely, the medical meaning of 'diagnosis' is not always, by definition, that of finding out or establishing precisely the nature of a disease known to be present in the client. There incontrovertibly is imprecise, uncertain, and even incorrect diagnosis, and not only about the nature of a known-to-be-present disease but also about the very existence a disease in the person in question at the time in question. And if indeed one of the meanings of the Greek *diagnōsis* is "a deciding," this is not instructive of the meaning of 'diagnosis' in medicine. The meaning of 'differential diagnosis' does need to be clarified. At issue naturally is diagnosis in the context in which the client presents with a sickness or an abnormal test result that is not pathognomonic for any single cause of this, so that more than one condition needs to be entertained for diagnosis.
2. Regarding that lucubration on the meaning of '*disease*,' then, a number of added critical points are in order. Starting from that purported literal meaning of the word, no dictionary or thesaurus of mine presents the word as an/the antonym of 'ease,' nor is absence of, say, coronary heart disease ever referred to as a type of 'ease.' It is circular to say that disease is a morbid entity, when morbid entity is defined as a "diseased" (or pathologic) entity; and regardless, no rational conception of an entity involves its being characterized "usually [*sic*] by at least two [*sic*]" of a set of criteria. The allusion to syndrome in this context is odd, especially as the meaning of this term is specified in reference to "any morbid process [*sic*]," with Down's syndrome – which actually is a pattern of clinical manifestations of a state (trisomy 21, say) rather than a process – presented as

one of them. As for that purported first meaning of 'disease,' the question is whether that remarkably pleonastic aggregate of words means anything other than, simply, somatic anomaly (cf. Sect. 5.1 above). Curiously, if the anomaly is not an "interruption" or a downright "cessation" of something (presumably a function), then – and only then – it is a "disorder" of that something, which is said to be a "disturbance" possibly resulting from "exogenous [*sic*] factors such as . . . disease [*sic*]." And that something, if it is not a "bodily function," then it is not a bodily structure but something else.

Medical dictionaries thus make two things very clear about such elementary, generic medical terms as 'disease' and 'diagnosis': their meanings remain very confused in today's medicine. And "authoritative" dictionaries of medicine compound the prevailing confusion about these most elementary concepts of medicine by their uncritical, confused, and mutually inconsistent ways of attempting to specify the purported meanings of the terms.

It is not "authoritative" medical dictionaries alone that manifest surprising misapprehensions about the centerpiece of medicine, diagnosis. They are amply on display in eminent textbooks of medicine, for example (Sects. 3.1 and 3.2).

At the root of these misapprehensions is the fact that quite universally, even among leaders of clinical medicine, it remains unappreciated that medicine is supposed to be an aggregate of *knowledge-based*, learned disciplines of practice; that in the practice of clinical medicine the doctor endeavors to achieve *esoteric knowing* about the health of the client; and that in this, the very first need commonly is to achieve diagnosis – knowing about the present state of the client's health (usually in respect to each of a particular set of illnesses that the client may have as the explanation of the chief complaint in the presentation of the case).

If these elementary principles of medicine (Sect. 1.1) were universally appreciated, there would be no tendency to conflate the concepts of diagnosis and the patient's illness, as when talking or writing about 'the patient's diagnosis' or about the illness in question as 'the suspected diagnosis.' As an elementary truism, the locus of possible illness is the soma of the patient, while that of diagnosis – about the patient's possible illness – is the mind of the doctor. Diagnosis in clinical medicine is a doctor's *knowing* (esoteric) about the presence/absence of a particular illness in the client (Sect. 5.2 above), a matter profoundly distinct from the reality this knowing is about.

This conflation of two profoundly different, though closely related, concepts is commonly manifest in textbooks of medicine in the context of 'differential diagnosis,' in presenting this as the set of illnesses – or even diagnoses – that could explain the patient's sickness (Sects. 3.1 and 3.2) instead of taking the term to refer to diagnosis about each of these (cf. Sect. 5.2 above).

Appreciation of those elementary principles would put an end also to the common parlance and writing about 'making a diagnosis.' Realistically, one does not 'make' something that is in the nature knowing. Instead, in diagnosis one brings general medical knowledge to bear on the facts constituting the diagnostic profile of the case, thus achieving diagnosis – esoteric knowing of that genre – about the case at hand (Sect. 1.1).

The misapprehensions surrounding diagnosis – this centerpiece of medicine – seem not to be clearing away due to conceptual progress in clinical medicine. Instead of such progress, *regress* appears to be taking place, exemplified by the recent emergence of the concept(s) of 'acute coronary syndrome' (Sects. 3.1 and 3.2) and the various theoretical ideas about diagnosis adduced by 'clinical epidemiologists' (Sect. 2.1, i.a.).

5.4   Diagnostic Tests: Misapprehensions of Type 1

My OED (1998) gives as the denotation of the noun '*test*' this:

> a procedure intended to establish the quality, performance, or reliability of something, especially before it is taken to widespread use.

Rephrased, this OED defines test as:

> a *procedure* producing information to establish the *quality* of something (on a defined scale of quality – as to performance or reliability, say).

And in less categorical terms, thus, a test is a procedure producing a *datum* (vector-valued rather than scalar, perhaps) for *inference* (uncertain perhaps) about the quality of something (on an a-priori scale of quality).

A *diagnostic test* is, of course, a particular type of test. Specifically, it is: a procedure producing a datum (vector-valued perhaps) for inference about the 'quality' of health as to presence/absence of a particular illness – for diagnosis about this, that is. It is a procedure that produces a datum for incorporation as an addition into the diagnostic profile of the case at hand.

Not all of the elements in a diagnostic profile are data derived by the application of a particular, test-defining procedure – those addressing the patient's demographics and symptoms most notably. But there can be *clinical tests* as part of the physical examination, and other types of clinical test – electrocardiography, for example – performed right in the doctor's office, to complete the clinical-diagnostic profile of a case. Upon the translation of this profile into clinical diagnoses, there commonly are *extra-clinical tests* to which the patient is referred when the diagnostic probability for a particular illness is of concern but is not extreme enough for action other than the testing. The question about testing a particular diagnostic possibility arises first when diagnosis first is the concern, namely once the clinical profile of the case has been completed as called for by the type of patient presentation. (Cf. Sect. 1.1.)

Two major research-distorting *misapprehensions* about the essence of a diagnostic test are of note, both of them serving to 'justify' a line of diagnostic research to which there is an unjustifiable a-priori commitment. These are, thus, not misapprehensions of doctors who use diagnostic tests' results for diagnosis but of researchers who have an a-priori commitment to a misguided line of diagnostic research.

One of these misapprehensions – let's call it Type 1 misapprehension – involves orientationally the idea that every item in a diagnostic profile is the result of a diagnostic test, the patient's age and chief complaint, for example (cf. Sect. 2.1). This notion is invoked to underpin the corollary of it that all of diagnostic research reduces to study of the *properties of diagnostic 'tests.'* And to this pair of root fallacies is added the misapprehension that each 'test' for the diagnosis about a particular illness has its relevant properties – ones of *'diagnostic accuracy'* – independently of what else is in the diagnostic profile – *singular values* for its 'sensitivity' and 'specificity,' or likelihood ratio for the 'test result' (the datum's likelihood/probability of occurrence conditional on the presence of the illness in question divided by the counterpart of this conditional on the absence of the illness; Sect. 2.1).

In the framework of these (false) premises the correct probability odds for the presence of a particular illness are taken to be

$$(\text{'pre} - \text{test' odds}) \times LR_1 \times LR_2 \times \ldots,$$

where the 'pre-test' odds are predicated on total absence of diagnosis-relevant facts about the case; LR_1 may be the likelihood ratio for the patient's age and LR_2 the counterpart of this for the patient's gender; LR_3 may address the chief complaint; etc. That calculation would be correct if that 'pre-test' odds were a sensible concept and those LRs actually were single-valued characteristics of the diagnostic indicators involved (Sect. 2.1).

To gain some orientational insight into the *fallacies* in this favorite aggregate of ideas of 'clinical epidemiologists,' let us consider two simple examples, involving only two diagnostic indicators, both binary in their realizations. And let us suppose that in the domain of the diagnosis (chief complaint and range of age, say) each of the two indicators give a positive result ($X_i = 1$) with probability 0.5, and that the prevalence/probability of the illness (I) in question in that domain is

$$Pr(I) = 0.04 + 0.16X_1 + 0.16X_2.$$

Finally for the *first example*, let us suppose that the distributions of the two diagnostic indicators are, in the domain in question, mutually independent, uncorrelated. (This is possible for two indicators of 'risk' for, but not for two manifestations of, the illness nor for a pair consisting from one of each.)

In this example, the 'pre-test' probability of I is $(0.04 + 0.20 + 0.20 + 0.36)/4 = 0.20$, and the 'pre-test' odds thus are $0.20/(1-0.20) = 1/4$. The 'sensitivity' of each of the two indicators (probability of $X_i = 1$ given presence of I) is $(0.20 + 0.36)/(0.04 + 0.20 + 0.20 + 0.36) = 0.70$, and the 'specificity' of each of them (probability of $X_i = 0$ given absence of I) is $[(1-0.04) + (1-0.20)]/[(1-0.04) + (1-0.20) + (1-0.20) + (1-0.36)] = 0.55$. The likelihood ratio for a positive result thus is $0.70/(1-0.55) = 1.56$, and for a negative result it is $(1-0.70)/0.55 = 0.55$. For $X_1 = X_2 = 0$ that LR-based calculation (above) gives $(1/4)/(0.55)^2 = 0.076$ instead of the correct $0.04/(1-0.04) = 0.042$. For $X_1 = 0$, $X_2 = 1$ and conversely, the LR-based result is $(1/4)(0.55)(1.56) = 0.21$ instead of the correct $0.20/(1-0.20) = 0.25$. And for $X_1 = X_2 = 1$ the result is $(1/4)(1.56)^2 = 0.61$ instead of the correct $0.36/(1-0.36) = 0.56$.

Let us now change one of the premises here, presuming in a *second example* that the two indicators are totally correlated, so that $Pr(X_1 = X_2 = 1) = 0.5$ and $Pr(X_1 = X_2 = 0) = 0.5$. The 'pre-test' odds naturally remain unchanged: $(0.04 + 0.36)/2 = 0.20$. 'Sensitivity' for each of the two indicators is now $0.36/(0.04 + 0.36) = 0.90$ instead of the 0.70 above, and the 'specificity' is $(1-0.04)/[(1-0.04) + (1-0.36)] = 0.60$ instead of the 0.55 above. Correspondingly, the LR values for $X_i = 1$ and $X_i = 0$ are $0.90/(1-0.60) = 2.25$ and $(1-0.90)/0.60 = 0.17$, respectively instead of the 1.56 and 0.55 above. For $X_1 = X_2 = 0$ the result for the LR-based odds of I thus are $(1/4)(0.17)^2 = 0.0072$ instead of the correct $0.04/(1-0.04) = 0.042$; and for $X_1 = X_2 = 1$ they are $(1/4)(2.25)^2 = 2.27$ instead of the correct $0.36/(1-0.36) = 0.56$.

The *theoretical flaw* in this treatment of a diagnostic profile in terms of item-specific likelihood ratios (in lieu of the multivariate function for the probability) lies in false generalization of the theoretically correct relation that

$$Pr(I)/[1 - Pr(I)] = (\text{Unconditional odds}) \times LR,$$

where the LR has to do with unidimentional (scalar) datum.

For a multivariate (vector-valued) datum the correct generalization of this involves

$$LR = \frac{Pr\left(X_1 = x_1, X_2 = x_2, \ldots | I\right)}{Pr\left(X_1 = x_2, X_2 = x_2, \ldots | \overline{I}\right)}.$$

This can correctly be decomposed as

$$LR = LR_1 \times LR'_2 \times LR'_3, \ldots,$$

where

$$LR'_2 = Pr\left(X_2 = x_2 | X_1 = x_1, I\right)/Pr\left(X_2 = x_2 | X_1 = x_1, \overline{I}\right),$$

LR'_3 is conditional on $X_1 = x_1$ and $X_2 = x_2$, etc.

The *flaw* at issue here is the failure to appreciate that LR'_2 is not the same as $LR_2 = Pr\left(X_2 = x_2 | I\right)/Pr\left(X_2 = x_2 | \overline{I}\right)$, etc.

As an *illustration* of this, let's think back to that latter example above. Whatever is the realization of X_1, X_2 replicates this. Therefore, with LR_1 accounted for, $LR'_2 = 1$ regardless of the realization of $(X_1$ and) X_2. Based on X_1 alone, then, the result with $X_1 = 0$ and $LR'_2 = 1$ is $(1/4)(0.17) = 0.043$, identical with the correct $0.04/(1-0.04)$, while with $X_1 = 1$ and $LR'_2 = 1$ it is $(1/4)(2.25) = 0.56$, identical with the correct $0.36/(1-0.36)$. The key to correctness is the use of LR_2 that is conditional on the value of X_1.

The theoretical flaw in the decomposition of the LR for a multivariate datum is of major consequence, notably because of the generally substantial correlations among the diagnostic indicators, the manifestational ones in particular. The main point is this: *a given diagnostic indicator has no innate 'accuracy'* for the diagnosis of a particular illness, quantifiable in terms of single-valued 'sensitivity' and 'specificity,' or likelihood ratios. In the second example above, the LR for $X = 1$ and $X = 0$ for both X_1 and X_2 are 2.25 and 0.17, respectively, when considered alone; but conditionally on the other X, both LR values are 1.00, signifying complete uninformativeness of the added, contextually wholly redundant test.

A diagnostic indicator's informativeness about the presence of a particular illness depends on what else is accounted for in the diagnostic profile; its informativeness is contextual, marginal and cannot be meaningfully addressed divorced from the context. In the examples above, each of the two indicators is addressed in the context of the other, and of the model $Pr(I) = 0.04 + 0.16X_1 + 0.16X_2$, which implies, for example, that when $X_1 = 0$, $Pr(I)$ is 0.04 or 0.20 according as $X_2 = 0$ or $X_2 = 1$, while if $X_1 = 1$, the corresponding X_2-determined probabilities of I being present are 0.20 and 0.36. In this example, X_2 implies, conditionally on X_1, a constant probability difference (and so does X_1 conditionally on X_2), but this is so only because the model is additive, instead of the more general $Pr(I) = B_0 + B_1X_1 + B_2X_2 + B_3X_3$, where $X_3 = X_1X_2$.

And in closing on this I note that as the concept of a diagnostic test's 'accuracy' is a seriously malformed one, no clinician should derive diagnostic probabilities by multiplying 'likelihood ratios' based on diagnostic indicators' presumedly context-independent values for these. But on a more cheerful note in this gloomy context, such LR-multiplying clinicians actually are very rare, if existent at all, as most clinicians remain quite unspoiled by 'clinical epidemiology,' the teachings about practice and research under this heading.

As for diagnostic *research*, however, the 'accuracy' of diagnostic 'tests' remains a very major preoccupation. Two articles are particularly illustrative of the leaders' thinking on this and of the extent of the followers' work along these lines (refs. 1, 2 below):

References:

1. Bossyt PM, Reitsma JB, Bruns DE et al (2003) Towards complete and accurate reporting of studies of diagnostic accuracy: the STARD initiative. Ann Intern Med 138:40–44
2. Bossyt PM Reitsma JB, Bruns DE et al (2003) The STARD statement for reporting studies of diagnostic accuracy: explanation and elaboration. Ann Intern Med 138:W1–W12

These authors, among many others, hold that "The term *test* refers to any method of obtaining additional information on a patient's health status. It includes information from history and physical examination, laboratory tests, imaging tests, function tests, and histopathology" (ref. 1 above). So, any method of "additional" inquiry into the chief complaint, for example, is a diagnostic test; and according to these authors, diagnostic laboratory tests do not include imaging tests, function tests, or histopathologic tests.

"At the 1999 Cochrane Colloquium meeting in Rome, The Cochrane Diagnostic and Screening Test Methods Working Group discussed the low methodological quality and substandard reporting of diagnostic test evaluations. The Working Group felt that the first step to correct these problems was to improve the quality of reporting of diagnostic studies. ... Complete and accurate reporting allows the reader to detect potential for bias in the study (internal validity) and to assess the generalizability of the results (external validity)." (Ref. 1 above.)

For "the development of a checklist of a study that should be included in the report of a study of diagnostic accuracy" there was a steering committee constituted mainly by 'clinical epidemiologists' and 'biostatisticians.' It conducted "an extensive search to identify publications on the conduct and reporting of diagnostic studies ... and contacted other experts in the field of diagnostic research. ... Subsequently, the STARD steering committee convened a 2-day consensus meeting for invited experts from the following interest groups: researchers, editors, methodologists, and professional organizations." (Ref. 1 above.)

"The search for published guidelines for diagnostic research [*sic*] yielded 33 lists. ... During the consensus meeting ... participants consolidated and eliminated items to form the 25-item checklist" (ref. 1 above). The items in this list run from #1: "Identify the article as a study of diagnostic accuracy (recommended MeSH heading 'sensitivity and specificity')" to #25: "Discuss the clinical applicability of the study findings." So, this was the only type of diagnostic research!

"We provide one prototypical flow chart that reflects the most commonly deployed design in diagnostic research" (ref. 1 above). In this type of study the 'test' in question provides a result that is either abnormal, normal, or inconclusive. The "reference standard" test is applied regardless, and its result is either "Target condition present" or "Target condition absent." That flow chart does not indicate the role of the data on the instances of inconclusive result of the 'test' under study, in deriving the results for its 'sensitivity' and 'specificity,' even though "The STARD group put considerable effort into the development of a flow diagram for diagnostic studies" (ref. 1 above).

Clarity on this puzzling matter can be gained by examining the most recent one of the very eminent PIOPED studies on the diagnosis of pulmonary embolism (refs. 3, 4 below). "The design of PIOPED III conforms to the design requirement [*sic*] of studies of diagnostic accuracy [refs. 1, 2 above] and to the Standards for Reporting of Diagnostic Accuracy (STARD) statements [refs. 1, 2 above]" (ref. 3 below). One of the tests was magnetic resonance angiography in combination with venography, the result of which was "technically inadequate in 52 % of patients (194 of 370)" (ref. 4 below). The "reference test" results were reported for these 194 as well as for the remaining 176 (cf. above) but then ignored: "Among patients with technically adequate images [176 of the 370] combined magnetic resonance angiography and venography had a sensitivity of 92 % ... (65 of 71) ... and a specificity of 96 % ... (101 of 105) ..." (ref. 4 below). In truth, however, the 'sensitivity' was only $65/104 = 63$ %, and the 'specificity' was only $101/266 = 38$ %.

References:

3. Stein PD, Gottschalk A, Sostman HD et al (2008) Methods of prospective investigation of pulmonary embolism diagnosis III (PIOPED III). Semin Nucl Med 38:462–470
4. Stein PD, Chenevert TL, Fowler SE et al (2010) Gadolinium-enhanced magnetic resonance angiography for pulmonary embolism: a multicenter prospective study (PIOPED III). Ann Intern Med 152:1–20

Item #18 in the check list (ref. 1 above) is this: "Report distribution of severity of disease (define criteria) in those with the target condition; other diagnosis in participants without the target condition." Comments are presented on the bearing of these matters on "measures of diagnostic accuracy," so that "it is important to describe the severity of disease in the study group" (ref. 2 above). But: nothing is said about the *correct mix of patients* for avoidance of "spectrum bias." The

PIOPED III investigators repeated the idea that "An important design requirement for a study of diagnostic accuracy is that the study should include a wide spectrum of the condition to be diagnosed [refs. 1, 2 above]" (ref. 3 above). "The study will include a broad spectrum of patients with and without PE and a variety of patients with comorbid conditions that are commonly associated with PE" (ref. 3 above). But: in the study report (ref. 4 above), nothing is said about the "spectrum" of patients (only the proportions with and without PE).

This August Working Group says nothing about the other topic in this framework of thought about diagnosis and diagnostic research, namely the '*pre-test*' *odds* (cf. above); but a very eminent group of 'clinical epidemiologists' does: "Can we generate a clinically sensible estimate of our patient's pre-test probability?" is a section heading in a book of theirs (ref. 5 below). The test under this heading begins thus:

> This is a key topic, and deserves its own 'section-within-a-section.' How can we estimate a patient's pre-test probability? We've used five different sources for this vital information: clinical experience, regional or national prevalence statistics, practice databases, the original report we used for deciding on the accuracy and importance of the test, and studies devoted specifically to determining pre-test probabilities.

Reference 5: Sackett DL, Straus SE, Richardson NS et al (2000) Evidence-based medicine: how to practice and teach EBM, 2nd ed. Churchill Livingstone, Edinburgh, p 82

So, implied by this is that a clinician concerned to know the probability that the patient complaining about chest pain has a case of ACHD (acute coronary heart disease) can perhaps draw the "vital information" about the "pre-test probability" of ACHD from some (undefined) clinical experience or from some (undefined) practice databases on the prevalence of ACHD; or (s)he can perhaps get it from national or regional routine data on this prevalence, again necessarily having to do with the aggregate of all cases, asymptomatic ones included; or there may have been surveys to expressly assess the level of this (single-valued) prevalence/probability, involving electrocardiographic and cardiac enzyme testings in representative samples of people, almost all of their members naturally free of any symptoms suggestive of the presence of ACHD.

The prevalence of ACHD in clinical experience and practice databases obviously is very highly variable among the various disciplines of medicine, so that these sources of the "vital information" on the "pre-test probability" of ACHD obviously are meaningless.

As for diagnosis-related surveys, it would be instructive to anyone under the Type 1 misapprehensions about diagnostic tests to see the results of a survey of expert emergency physicians regarding their "vital information" on the "pre-test probability" of ACHD, or the counterparts of this for pulmonary embolism, for example.

Most of them presumably would never have even thought about that probability; and if they'd hazard to guess, the values for this "vital information" presumably could vary by a full order of magnitude.

Were this type of survey to be conducted, those who still would believe in the vital importance of that "pre-test probability" presumably would be interested in a follow-up survey to ascertain what then is, in the thinking and practices of diagnostic experts, the alternative type of starting value for the odds ratio that is multiplied by the likelihood ratios for each of the test results (that for age, etc.) to arrive at the final "post-test odds" and then "post-test probability" for the illness in question. And it would be instructive for them to learn that this LR multiplication, just as the "pre-test odds" and "pre-test probability," are not at all part of their thinking; that they focus on the *diagnostic profile as a whole* and judge the probability for the illness is question on the basis of this as such and alone – without thinking about the LR for the whole profile together with the pre-profile probability or odds for the illness.

It deserves note that multiplication of LRs did not come up in textbook descriptions of the diagnoses about ACHD and PE reviewed in Sects. 3.1 and 3.2.

5.5 Diagnostic Test: Misapprehensions of Type 2

Recall from Sect. 5.4 above this: "The term *test* refers to any method of obtaining additional information on a patient's health status. It includes information from history and physical examination, . . ." (ref. 1 there). So, those under misapprehensions of Type 1 have their very first idea right: a diagnostic test produces an item (multi-dimensional perhaps) of *information* about the client's health – present health, to be specific. A procedure to produce a piece of information is not a procedure or action to intentionally produce a change in anything except in the state of available information; it is *not an intervention* (to change some externality for the better).

In line with this, my *OED* (1998) specifies the denotation of 'intervention' as:

action taken to improve a situation, especially a medical disorder: *two patients were referred for a surgical intervention.* [Italics in the original.]

And similarly, my *Dorland's* (1994) gives, following "[L. *intervenire* to come between]," this as the concept of intervention:

1. The act or fact of interfering so as to modify.
2. Specifically, any measure whose purpose is to improve health or to alter the course of a disease.

One more item of background here: while the desired, useful property of a diagnostic test is its *informativeness* about the presence/absence of a particular illness in the client, for a medical intervention the counterpart of this is its *efficacy* (or effectiveness).

According to my OED the denotation of 'efficacy' is:

the ability to produce a desired or intended result: *there is little information on the efficacy of this treatment.* [Italics in the original.]

And my Dorland's, following "[L. *efficax* effectual]," gives this:

1. The ability of an intervention to produce the desired beneficial effect in expert hands and under ideal circumstances [*sic*]. Cf. *effectiveness*.
2. In pharmacology, the ability of a drug to produce the desired therapeutic effect; it is independent of *potency*, which expresses the amount of the drug necessary to achieve the desired effect.

These elementary concepts of medicine (setting aside that purported, recently contrived efficacy vs. effectiveness duality) were first upended by radiologists and epidemiologists concerned with screening for a cancer, in which they mistook community-level *application of a screening test* (mammography) to be an *intervention* to reduce mortality from a cancer. This misapprehension led to the corollary misconception that the intended consequence of screening for a cancer is best assessed, like that of any medical intervention, by means of *randomized trials*. The most recent and most spectacular example of these misapprehensions in cancer diagnostics is the National Lung Screening Trial, contrasting CT screening for lung cancer with conventional radiographic screening for it and "concluding" (per $P < \alpha$) that "screening with low-dose CT reduces mortality from lung cancer" (example 7 in Sect. 7.5). Earlier, some 650 thousand women had been involved in randomized trials on the "efficacy" of mammography, which nevertheless remains a matter of heated controversies.

As for diagnostics outside screening, the National Cancer Institute of the U.S. in 1997 adduced a call for Cooperative Trials in Diagnostic Imaging (ref. 1 below), in which it said this:

more accurate images by themselves will not necessarily motivate new equipment purchases without evidence that the greater accuracy will translate into cost savings or better clinical results. These kinds of endpoints are most persuasively assessed using *rigorous clinical trials* methodology.... Where appropriate, this evaluation should include estimates of the relative cost- effectiveness of *diagnostic interventions* and their impact on quality of life. [Italics added.]

Reference 1: NIH guide (1997) vol. 26, 22 Aug

The upshot of this call was the formation of the American College of Radiology Imaging Network. "The specific objectives of ACRIN include ... [assessment of] imaging technologies beyond the evaluation of accuracy to include such end points as the *effect of imaging* examinations on medical diagnosis, treatment, and health care outcomes, including quality of life and health care costs" (ref. 2 below). (Italics added.)

Reference 2: Hillman BJ, Gatsonis C, Sullivan DC (1999) American college of radiology imaging network: new national collaborative group for conducting clinical trials of medical imaging technologies. Radiol 213:641–645

In the background of these developments was an earlier article by two radiologists (ref. 3 below). It distinguished among *six "levels of efficacy"* for a diagnostic imaging test:

1. Technical efficacy [e.g., "resolution of line pairs"]
2. Diagnostic accuracy efficacy [e.g., "sensitivity and specificity in a defined clinical problem setting"]
3. Diagnostic thinking efficacy [e.g., "difference in clinician's subjectively estimated diagnosis probabilities pre- to post-test information"]
4. Therapeutic efficacy [e.g., "number of [*sic*] percentage of times clinicians' prospectively stated therapeutic choices changed after test information"]
5. Patient outcome efficacy [e.g., "percentage of patients improved with test compared with without test"]
6. Societal efficacy [e.g., "cost-effectiveness analysis from societal viewpoint"].

Reference 3: Fryback DG, Thornburry JR (1991) The efficacy of diagnostic imaging. Med Decis Making 11:88–95

This diagnostic thinking of Fryback and Thornbury remains well-respected by those in the interface of radiology and health policy. One recent example of this is the thinking of The Working Group on Comparative Effectiveness Research (ref. 4 below): "Each of the levels outlined by Fryback and Thornbury can be used as a guide for making decisions about the extent of research needed for a particular technology. . . . It is important to note that, ideally, we would prefer all technologies to be held to the *highest standards of evaluation*, such as *randomized controlled trials.*" (Italics added.)

Reference 4: Gazelle GS, Kessler L, Lee DW et al (2011) A framework for assessing the value of diagnostic imaging in the era of comparative effectiveness research. Radiol 261:692–698

One example of those "highest standards of evaluation" is the *CECaT trial* (ref. 5 below). "The trial participants were patients with suspected or known CAD [coronary artery disease] and an exercise test result that required non-urgent angiography. . . . Patients were randomized to one of four initial diagnostic tests [as "interventions"]: angiography (controls); single photon emission computed tomography (SPECT); MRI; stress echocardiography. . . . The main outcome measurements were as follows: Primary: at 18 months post-randomization: exercise time (modified Bruce protocol); cost-effectiveness compared with angiography (diagnosis, treatment and follow-up costs). . . . At 18 months, comparing SPECT and stress echo with angiography, a clinically significant difference in total exercise time can be ruled out [while as for MRI] a difference of at least 1 min cannot be ruled out. . . . Cost-effectiveness was mainly influenced by test costs" ["with very little difference in quality-adjusted life years"].

Reference 5: Sharples L, Hughes V, Crean A et al (2007) Cost effectiveness of functional cardiac testing in the diagnosis and management of coronary artery disease: a randomized controlled trial. The CECat trial. Health Technol Assess 11(49):iii–iv

A quite closely related, second and final example here is *ROMICAT II* (ref. 6 below). It "was a randomized comparative effectiveness trial enrolling patients 40–74 years old without known coronary artery disease who presented to the ED [emergency department] with chest pain but without ischemic electrocardiographic (ECG) changes or elevated initial troponin and who required further risk stratification. Overall, 1000 patients at 9 sites within the United States were randomized to either CCTA [cardiac computed tomographic angiography] as the first diagnostic test following serial biomarkers or to standard of care, which included … Test results were provided to ED physicians, yet *patient management was not driven by a study protocol in either arm.* Data were collected on diagnostic testing, cardiac events, and cost of medical care for the index hospitalization and during the following 28 days. The primary end point was length of hospital stay. Secondary end points were … Tertiary end points were … Rate of missed acute coronary syndrome within 28 days was the safety end point. The ROMCAT II *will provide rigorous data* on whether CCTA is more efficient [*sic*] than standard care in the management of patients with acute chest pain, at intermediate risk for acute coronary syndrome." (Italics added.)

Reference 6: Hoffman U, Truong QA, Fleg JL et al (2012) Design of the rule out Myocardial Ischemia/Infarction using computer assisted tomography: a multicenter randomized comparative effectiveness trial of cardiac computed tomography versus alternative triage strategies in patients with acute chest pain in the emergency department. Am Heart J 163:330–338

These Type 2 misapprehensions about diagnostic tests can be shown to be misapprehensions by applying "rigorous clinical trials methodology" (ref. 1 above) in the assessment of the "efficacy" of some highly informative diagnostic test in improving the course of the patients' health. A 'thought experiment' will suffice: Patients are randomly allocated to the actual (verum) test in question or to a sham alternative to this in a double-blind manner, so that neither the patients nor the investigators know to which arm of the trial any given patient belongs, and in the instances of the verum test they thus remain blind to the test result. The experience in the trial shows that, different from an effective intervention, application of even a highly informative diagnostic test does not, in itself, "translate into cost savings or better clinical results" (ref. 1 above). And the reason is obvious to anyone not under the spell of the Type 2 misapprehensions about diagnostic tests: a test is not an intervention (cf. above).

Thus, the ACRIN has an empty mission (ref. 2 above); Fryback and Thornbury had misapprehensions on six levels (ref. 3 above); these misapprehensions will mislead, on various levels, the "comparative effectiveness research" on "the value of diagnostic imaging" (ref. 4 above); the CECaT trial, despite its 115-page report (ref. 5 above), did not make a contribution to the knowledge-base of diagnosis about coronary artery disease; and the "rigorous data" from the ROMICAT II trial on "whether CCTA is more efficient [*sic*] than standard care in the management of patients with chest pain at intermediate risk for acute coronary syndrome" (ref. 6 above) will not advance the knowledge-base of diagnosis about acute myocardial ischemia (acute coronary heart disease).

All six of the articles above are, and the only real-study example is Sect. 5.4 above (refs. 3, 4 there) also is, very notably, about diagnostic and diagnostic-research ideas by *radiologists*; and this is so because they are, quite uniquely among the disciplines of clinical medicine, the propagators of both types of misapprehension (Type 1 in Sect. 5.4 above, Type 2 here) about diagnostic testing and about research on diagnostic tests. This, in turn, likely is so because radiologists are prone to see themselves as diagnosticians par excellence, as they, uniquely, are able to 'see through' into the inner structures of the patient's body by means of the magic of their X-rays; because, for this reason, they are prone to have a distorted view of diagnosis and of their own role in the diagnostic pursuit. They have difficulty seeing the proper role of diagnostic radiologists to be akin to that of clinical chemists, for example – not that of diagnosticians but merely that of contributors of diagnostic data (from a laboratory) to supplement the pre-radiology elements in the patient's diagnostic profile, for diagnosis by the patient's actual doctor (at the patient's side rather than in a laboratory).

Following the delineation of some major misapprehensions about diagnosis and diagnostic research in Sects. 5.1, 5.2, 5.3, 5.4 and 5.5 above, I now continue unencumbered by these misapprehensions. In the three sections immediately following I address diagnostic research at large (Sects. 5.6, 5.7 and 5.8) and research specific to a diagnostic test I address in the section following these (Sect. 5.9).

5.6   A Paradigmatic Diagnostic Study

Diagnosis about "acute ischemic heart disease," constituted by "acute myocardial infarction, unstable pre-existing angina pectoris, and new-onset angina pectoris," was the concern in what I regard as the pioneering paradigm for diagnostic research (ref. below).

Reference: Pozen MW, D'Agostino RB, Mitchell JB et al (1980) The usefulness of a predictive instrument to reduce inappropriate admissions to the coronary care unit. Ann Intern Med 92:238–242

The authors developed a *probability function* for diagnosis about this AIHD entity for application in emergency-room triage decisions about referral of patients into the coronary care unit. In the second stage of the study they assessed the consequences of the availability of the probabilities derived from this function on the decisions about CCU referral.

Pozen et alii specified "105 variables that were both related to [AIHD] and available to the physician in the emergency room. These variables included clinical presentations, past history, sociodemographic characteristics, risk factors, physical findings, and electrocardiograms (ECGs)." They collected the data on these "variables" on "925 consecutive patients seen in [an] emergency room for suspected [AIHD]."

They fitted a logistic model (Sect. 2.1) to the data on all of the variates and then reduced it to retain only nine of them. "In decreasing order of statistical significance" the retained nine variates had to do with these binary, all-or-none features of the cases:

[1] history of previous myocardial infarction,
[2] T-waves if abnormal and inverted 1 mm or more or elevated more than 25 % of the height of the R-wave,
[3] ST segments [abnormal],
[4] chest pain located in lower or midsternum,
[5] patient reports of chest pain as most important symptom,
[6] history of angina pectoris,
[7] ST segments ... abnormal and depressed 1 mm or more or elevated 1 mm or more, and
[8] T-waves ... abnormal.

The respective variates (numerical) were defined as $X_i = 1$ if the ith feature is present, 2 (*sic*) otherwise. The common definition of an indicator variate, however, involves '1 if present, 0 otherwise' (cf. Sect. 2.1). In these latter terms the result for the diagnostic probability function was of the form

$$\Pr(\text{AIHD}) = 1 / \left\{ 1 + \exp\left[-\left(B_0 + \sum_1^9 B_i X_i \right) \right] \right\},$$

with the empirical values of B_0 trough B_9, respectively, -5.29, 1.56, 1.61, 1.10, 1.68, 1.58, 0.96, 0.86, 1.41, and 0.45.

When all of those features are absent (i.e., when $X_i = 0$ for all i), this empirical function gives the minimum value for the diagnostic probability. It is 0.5 %. And when all of those features are present (i.e., when $X_i = 1$ for all i), the probability estimate from this function is 99.7 %. For a case with normal ECG (i.e., $X_2 = X_4 = X_8 = X_9 = 0$) the estimate is never higher than 68 %; and if, in addition, there also is no substernal chest pain as the most important symptom (i.e., also $X_5 = X_6 = 0$), the maximum result is 15 %.

The authors were explicit about admissibility into the series of "925 consecutive patients seen in [an] emergency room for suspected [AIHD]" (cf. above). The suspicion was based on two criteria: age 30 years or more for men, 40 years or more for women; and the complaint of at least one of "chest pain; difficulty in breathing; upper abdominal pain, nausea, or both; fainting, dizziness, or both; palpitations; pedal edema; unexplained tiredness, weakness, or both; unexplained irritability; and pain in arms, shoulders, neck, throat, or any of these." A further requirement for admissibility naturally was informed consent.

Its various deficiencies notwithstanding, paradigmatic about this study is its focus, over three decades ago already, on a designed diagnostic probability function for a designed domain of this, and freedom from preoccupation with any purported 'accuracy' properties of diagnostic 'tests' such as asking about the symptoms or the history of angina pectoris. (Cf. Sects. 2.1 and 5.4.) More on this study below.

5.7 Study of a Diagnostic Indicator Set: Objects Design

In reference to the *domain* of admissibility into their study (by age and type of complaint), Pozen et alii designed a logistic *model* for the probability/prevalence of AIHD (Sect. 5.6 above). The parameters in that model – $B_0, B_1 \ldots$ (Sect. 2.1) – were the designed particular *objects* of their study, the more proximal generic object having been the set of diagnostic indicators, specifically that set's informativeness about the presence/absence of AIHD in the domain of the study. By defining subdomains within the domain of the study, the set of diagnostic indicators served to discriminate between the presence and the absence of AIHD, the set's various realizations (as diagnostic profiles of particular cases) implying case-specific probabilities for the presence of AIHD.

The designed statistical model for the probability/prevalence of AIHD in the domain of the study defined the terms in which the investigators elected to think about that probability in that domain, thereby defining the particular objects of the study. For the great majority of those object parameters the study result was $B_i = 0$, in the meaning of no statistically significant deviation from this, with the non-zero results (statistically significant in their differences from zero) specified in Sect. 5.6 above.

The *design* of the objects of a diagnostic study begins with that of the *domain* of its object function (to be designed in reference to its predesigned domain). Pozen et alii designed the domain very inclusively, remarking that "more than 90% of all patients with [AIHD] who seek medical care" are from the domain of their study. Alternatives to this would have been, first, a single study for a particular 'chief complaint,' chest pain, for example; and second, separate studies for each of more than one chief complaint – with the domain for each defined by gender and age in addition to the single chief complaint.

In contemplating these alternatives it is important to bear in mind the role of the chief complaint in actual diagnostic work-up of a case: it, in conjunction with the rest of the domain definition, determines what facts are to be ascertained for the first-stage, clinical diagnostic profile, starting from particulars of the chief complaint itself (Sect. 1.2); and in particular, this domain definition implies the differential-diagnostic set of possible causes of the sickness and hence the set of diagnostic indicators to consider for discrimination among these (Sect. 2.1).

In terms of the thinking of Pozen et alii, one subdomain for diagnsosis about AIHD involves "chest pain" as the chief complaint, another one involves "difficulty in breathing" in this role, etc., all the way down to "pain in arms, shoulders, neck, throat, or any of these" (cf. above). But among the respective subdomains, the suitably designed forms of the respective probability functions naturally are very different in respect to the descriptors of the chief complaint; and they may well be different in some other respects as well, as the differential-diagnostic set is generally specific to any given chief complaint (in a defined demographic category).

So the *central principle* in the design of the domain for diagnostic probability function is this: A properly-designed domain for a DPF does not cut across the various chief complaints in the context of which the illness at issue is a member of the differential-diagnostic set, with the aim of developing a shared DPF for various suspicion-raising chief complaints. For, as an example, the correct DPFs specific to each of the complaints involved in the domain of the Pozen et alii study would likely be very different; and in the result's application, only a particular chief complaint is involved. A properly-designed domain for a DPF is, thus, specific for a particular generic type of chief complaint, with the particulars of this accounted for in the DPF (along with particulars of the demographic aspect of the domain). While the Pozen et alii study is paradigmatic in its pioneering development of a DPF, the same is not true of its design of the domain for this DPF.

As for the design of (the form of) the *object function* itself, let us continue to address the diagnosis about AIHD, focusing on the domain of *chest pain* as the chief complaint by a person from the same demographic category as in the Pozen et alii study.

In this design the point of departure is the domain-specific *differential-diagnostic set*: apart from AIHD, chest pain in that demographic domain can be a symptom of pneumonia, pulmonary embolism, and pericarditis, among other possibilities. For, the designed set of diagnostic indicators for the diagnosis about AIHD in this domain of chief complaint and demographics is to be thought of as serving discrimination between AIHD and some unspecified one of these particular alternatives to it – and not between AIHD and some wholly unspecified alternative – in the causation of chest pain. Thus, age does not serve as an indicator for the diagnosis about AIHD simply on the ground that it is an indicator of the risk (momentary) for AIHD; for it to be a diagnostic indicator in this context requires that it would serve as a discriminator between AIHD and that particular set of alternatives to it in the causation of chest pain (in the demographic category that is co-definitional to the domain). The same applies to whichever indicator of risk and also to whichever manifestation of illness, particulars of chest pain and ECG findings included. A feature relatively peculiar to one (or more) of the alternatives takes away from the probability of AIHD.

Given an inclusive set of potential diagnostic indicators for inclusion in the design of (the form of) the DPF for the designed domain for this, the question of potential reduction of this set arises. An unduly large set has the drawback that studying the corresponding DPF requires an unduly large series (of instances from the DPFs domain). The need is to keep the number of parameters down to a suitably small proportion of the number of instances (person-moments) constituting the study series multiplied by $P(1-P)$, where P is the proportion of cases of AIHD in the series. A suitable aim would be to keep this proportion down to 5 % or less, in part by parsimony in the design of the (form of the) DPF.

In this spirit of *parsimony*, carried to its extreme, the designed DPF might have been a logistic model involving only six independent variates:

X_1: indicator of positive history for AIHD;
X_2: indicator of sickness (specified type of chest pain, etc.) strongly suggestive of AIHD;
X_3: indicator of sickness (specified) moderately suggestive of AIHD;
X_4: indicator of ECG changes (specified) strongly suggestive of AIHD;
X_5: indicator of ECG changes (specified) moderately suggestive of AIHD; and
X_6: indicator of ECG changes (specified) weakly suggestive of AIHD.

Explicit definition is needed for each of the six facts indicated by $X_i = 1$ (rather than 0). ($X_2 = X_3 = 0$ implies sickness only weakly suggestive of AIHD, and $X_4 = X_5 = X_6 = 0$ implies ECG pattern with no suggestion of AIHD.)

The parsimony of this model, involving only seven parameters, is achieved by careful definition of the multifaceted facts indicated by $X_2 = 1$ though $X_6 = 1$ in particular. The principle here is avoidance of essentially vacuous distinction-making together with maximal exploitation of clinical expertise for the definition of those composite entities indicated by $X_i = 1, i = 2, 3, \ldots, 6$, for this reduction of the realizations of a large number of case descriptors accounted for in those of a small number of simple statistical variates.

This simple DPF assigns the lowest probability of AIHD to instances of its demographic domain with weakly suggestive sickness alone pointing to AIHD, the highest probability to instances with positive history of AIHD as well as both the sickness and the ECG strongly suggestive of AIHD. It distinguishes among 24 types of case, 22 intermediate between those extremes translating to (practical) rule-out and rule-in probabilities.

The development of a parsimonious model like this – one in which the number of parameters is kept to a bare minimum – requires very considerable reliance on and deployment of pre-existing diagnostic expertise about the domain of the study. For needed is a-priori construction of ways to summarize the various features of the patient's sickness and the various findings from the ECG into a pair of corresponding ordinal degrees of suggestiveness about the presence of AIHD. This could perhaps be accomplished rather realistically by cultivating expertise as described in Sect. 9.1. Given these a-prioristic ordinal scales, the point of the actual study would be to translate the joint realizations on these scales into research-based empirical values for the diagnostic probabilities.

An alternative to the ordinal indicators' representation as in the model above (by indicators variates) is to translate them into quasi-quantitative indicators, with suitable allowance for curvature in the probability logit's dependence on the score values.

The definition of those two categories for the type of sickness is particularly challenging in the design of this model, which has the virtue of extreme parsimony conditionally on those definitions. As a simple model for the diagnosis of MI was presented in Sect. 2.1, the point was made that it was unrealistic on the basis of not involving consideration of the nature of the sickness in the presence of the differential-diagnostic alternatives of MI. Said was that, "Whereas that model has an indicator for the [chest pain] being more-or-less typical of MI (substernal 'aching' or 'pressure'), it should, in the same vein, have an indicator for the [chest pain] being more-or-less typical of gastroesophageal reflux (substernal 'burning'), for example." Here the challenge is to *account for the other illnesses* in the design of the two MI-oriented categories of sickness characterizing and surrounding acute chest pain.

The essence of this challenge is the need to develop, a-priori, a scheme of summarizing the features of the patient's sickness into a unidimensional, ordinal score of MI-suggestiveness. In it, each feature has a numerical value, positive or negative, multiplied by a weighting factor, and the score is the sum of these elements. The negative sign for a given element represents the idea that the feature is more typical of an alternative to MI than to MI. Given such a scoring function, ranges representing strong and moderate suggestiveness (re X_2 and X_3) can be defined.

5.8   Study of a Diagnostic Indicator Set: Methods Design

Turning from objects design to methods design, let's continue with that example in Sect. 5.7 above, representing an outgrowth of the paradigmatic diagnostic study by Pozen et alii, introduced in Sect. 5.6.

As in epidemiological research on the etiogenesis of an illness, the very first concern in this study's methods design is the selection of a suitable *source population*, here for the purpose of identification of instances of the study domain (defined by demographics and chief complaint). And as in the Pozen et alii study, the concern is to have a source of cases of the sickness early in its course, as diagnoses in these bear on emergency decisions about referral/non-referral to a coronary care unit. So, as in the Pozen et alii study, a suitable source population is formed by entries into a hospital's emergency department, or several of these EDs.

Pozen et alii enrolled successive instances of the study domain from entries into an ED, following a routine principle of clinical research within the constraint of the patient's informed consent. The general idea in this is that a study's validity requires a representative series of cases, and that the enrollment of successive cases serves to assure representativeness. This principle would be appropriate, and indeed important to heed, if the aim of the study were to describe the distributions of the various diagnostic indicators and/or their composite profiles of patients with AIHD; but it has no relevance for a study of the prevalence of AIHD as a function of the diagnostic indicators.

The way the prevalence of AIHD depends on the set of independent variates in the logistic model above (Sect. 5.7) – specified by the set of values of the object parameters B_0 through B_6 – is prone to be independent of the distribution of the independent variates in the study series of instances of the study domain. By design or otherwise, one study may accent low-probability cases and another one cases with relatively high probabilities of AIHD. But so long as both of them cover the entire set of indicator-realization patterns that actually occur in the function's domain, the resulting probability functions do not reflect this duality; their differences are 'due to chance' alone.

Within that framework of case coverage, *validity of case selection* requires only that inclusion in the study series is independent of the actual presence/absence of AIHD conditionally on the set of X realizations, which is what the study is about; and this independence is assured by enrollments before the truth about the AIHD status is known together with the enrollments being independent of indicators of this AIHD status not accounted for in the designed object function.

Given that X-selectivity of enrollment commonly is fully consistent with the study's validity, such selectivity could be employed to the end of optimizing the study's *efficiency*, that is, minimizing its 'cost' (monetary and other) for a given level of its result's precision (reproducibility). The source population should not determine the distribution by the Xs in the study series. Deliberate X-selectivity of enrollments would have as its aim increased variability and decreased correlatedness of the Xs in the study series. In this particular study, however, this selectivity may not be called for, as determination of the truth about the presence/absence of AIHD is not an imposition on the patient nor particularly costly to the investigators.

Enrollments' selectivity does not serve the study's efficiency alone; it is needed for *validity* assurance as well. Enrollment is consistent with validity only insofar as it allows valid documentation of the diagnostic profile and secure determination of the truth about the AIHD status. For admissibility of an instance from the study domain, the patient must be able to give an accurate and intelligible description of the sickness, and subsequent determination of the AIHD status must be possible. If there is doubt about either one of these two requirements for valid contribution, the instance is ineligible for inclusion in the study series.

With the data collected and translated into the realizations of the statistical variates in the model, the model is fitted to the data, to obtain the empirical values ('estimates') of the parameters together with values for the 'standard errors' of these. Model reduction on the basis of statistical significance (à la Pozen et alii) is not justifiable, nor is it even a temptation in the context of a model suitably designed to avoid overparametrization (Sect. 5.7 above).

5.9 Study of a Diagnostic Test: Objects Design

Further to that example, on the premise that the study size is large enough for reasonable validity of the result (for freedom from material 'overfitting,' really meaning undersizing), the function's application in a given case will commonly give a probability estimate not extreme enough for either rule-out or rule-in diagnosis about AIHD in the operational sense of justifying either reassurance of absence of AIHD or referral to the CCU.

This situation in the ED raises the question of whether the application of a cardiac enzyme test, perhaps one of the troponin tests in particular, would provide a result such that the corresponding post-test probability estimate would be sufficiently extreme for one of those possible actions. Actually, the question is not about whether, but about the *probability* with which, the post-test probability estimate would turn out to be extreme enough to be action-justifying. The diagnostician needs a knowledge-base for the estimation of that probability, for the decision about invoking the test.

When an AIHD-diagnostic study is to suitably inform the decision about the invocation of one of the troponin tests to supplement the routine ECG test, the first-order need is to study two sets of diagnostic indicators, the pre-test set and the post-test set. The pre-test probability function in Sect. 5.7 needs to be supplemented by a corresponding *post-test function*, its form just as carefully designed.

In the design of (the form of) this post-test function the beginning naturally is introduction of X_7: troponin level (in plain numerical terms). But as the (behavior and) meaning of this test result is strongly dependent on the pre-test probability, allowance for this is needed. One possible, simple way would be to fit the post-test model involving the linear compound $B'_0 + B'_1 L_6 + B'_2 X_7 + B'_3 L_6 X_7$, where L_6 is the value of the linear compound in the pre-test function, of the logit of the pre-test probability.

This post-test function allows the diagnostician to determine which range(s) of the test result, if any, would translate into a sufficiently conclusive post-test probability, given the pre-test profile of the case. Insofar as there is such a range, the need is to consult a function that gives (an estimate of) the probability of the test's result falling in this range, given the pre-test profile of the case.

Thus the need in the diagnostic research in respect to AIHD is to supplement the pre- and post-test functions as objects of study by functions (logistic) for the probabilities of various ranges of the troponin result – possible rule-out or rule-in ranges – as these probabilities depend on the variates describing the pre-test profiles of the cases. For any given one of these ranges of test result, the logit of its probability could replace the logit of the probability of AIHD in the pre-test model for the latter.

Adding these test-focused objects to diagnostic studies brings no notable novelty to the methods of data collection in these studies (other than adding the troponin test), nor really to fitting of any given model to the data (suitably translated into the realizations of Y, X_1, etc.).

As for the objects, though, it needs to be added that the test result need not be unidimentional, as in the case of the troponin test. CT coronary angiography, for example, produces a result with facts in more than just a single dimension. So there will have to be more than a single X representing such a test's result in the model; but this may well be so even in respect to a unidimensional test result (e.g., when providing for quadratic curvature in the probability logit's relation to the test result). In general, then, the test result in the post-test function is represented by the sum of a set of $B_i X_i$ terms pertaining to it, and this sum/score has a range implying the range of post-test probabilities conditional on the pre-test profile, etc., as outlined for a single term above.

5.10   Some Recent Studies Critically Examined

Recent reports on original studies for diagnosis I examine in terms of those appearing in the 2012 volumes of *JAMA*. There were only two of them.

Study 1 was this:

Reference 1: Berg WA, Zhang Z, Lehrer D et al (2013) Detection of breast cancer with addition of annual screening ultrasound or a single screening MRI to mammography in women with elevated breast cancer risk. J Am Med Assoc 307:1394–404

This study had been approved not only by the institutional review boards of all of the 21 participating institutions but also by the National Cancer Institute Cancer Imaging Program. It thus is well illustrative of the prevailing culture of an important genre of diagnostic research, that for the knowledge-base of diagnosis about a latent case of a cancer.

This study's Objective is said to have been "To determine supplemental cancer detection yield of ultrasound and MRI in women at elevated risk for breast cancer." Realistically, however, the objective could only have been to produce *evidence* on the *object* of study – on the magnitudes of the parameters in it – the object being a diagnostic probability function. For, no study is able to "determine" the magnitudes of the parameters in a DPF. (Cf. Sect. 5.3.)

A latent case of breast cancer is not detected (rule-in diagnosed) by means of 'mammography' alone, nor by this radiography supplemented by ultrasound imaging or MRI. It is detectable only by biopsy supplementing such imaging – by positive pathology result on the biopsy specimen following positive result of the imaging. It thus is detected – potentially – by an imaging-cum-biopsy *algorithm*; and implementation of this algorithm, rather than of the initial test alone, should be taken to be *the concept of the screening*.

In regard to a radiography-cum-biopsy algorithm, any appropriate DPF addresses the probability of the detection (rather than presence) of the cancer as a joint function of a suitable set of diagnostic indicators for a suitably defined domain of the algorithm's potential application. There is little point in restricting the domain to "women with elevated breast cancer risk" when, as is necessary, indicators of the level of the "risk" – meaning of the level of the probability at issue, with measures of breast density among these – are accounted for by the DPF; but a distinction needs to be made between the domains of baseline screening and periodic repeat screening.

With such a radiography-related DPF as the contextual given, the essential issue in the *objects design* in regard to diagnostic yield from supplementary ultrasound imaging in the algorithm is the way in which the form of that DPF is to be expanded to account for the marginal information from the added test. A reasonable question for the expansion to address would be: How-many-fold does the supplementary testing make the probability of detection of the cancer?

To address this question, the DPF defining the context should be log-linear rather than logistic; or if it is logistic, it should be thought of as being log-linear (which is realistic, given that the probabilities are very small). The minimally needed expansion of this model is very simple: mere addition of a term for an indicator of the supplementary testing. The exponential of the coefficient of this indicator is the factor by which the detection probability is increased, the proportional increase being this factor minus one.

The needed extension of this model to address screening that involves MRI as a supplement to radiographic-cum-ultrasound imaging is obvious, and so is the interpretation of the coefficient of the MRI indicator in it.

To be understood in all of this is that the ultrasound test is indicated (according to the algorithm) only if the result of radiography is negative (as defined by the algorithm); that the MRI test is indicated only if the result from both of the first two imaging tests is negative; and that biopsy is indicated if, and only if, the result is positive from the radiography as the only imaging test or, in general, from the last one of the imaging tests.

The essence of the requisite *methods design* flows from these understandings of the essentials of its objects design. Let us first focus on a study on the yield implications of adding the ultrasound test to the diagnostic algorithm at baseline.

The abstract of the report could read like this: Object: the proportional increase in the diagnostic yield of radiography-cum-biopsy baseline screening for breast cancer from adding ultrasound testing to the algorithm. It thus was the coefficient of the indicator of that supplementary imaging in a log-linear model designed for the diagnostic probability. Methods: The designed screening algorithm, addressed by the model, was applied to women from the model's domain, and the model was fitted to its variates' realizations in the data on them. Result: The rate of cancer detection was . . . % higher on account of the ultrasound testing upon negative result of mammography. The 95 % imprecision interval for this result was . . .% to . . .%.

The counterpart of this for a round of periodic repeat screening is completely analogous to this, and so also are the counterparts of both of these when the screening algorithm involves MRI upon negative result from both radiographic and ultrasound imagings.

With this sketch of what the nature of the first example study should have been as the background, I turn to study 1 as it actually was. The contrast is very sharp, in terms of the big picture already.

The 18 authors described a study that had a clinicaltrials.gov identifier – a study that was a randomized trial, that is, as though an intervention-prognostic study was the appropriate means "To determine supplemental cancer detection yield of ultrasound and MRI in women at elevated risk for breast cancer." In this spirit, in the main part of the study, the participants were scheduled to undergo "3 annual independent screens with mammography and ultrasound in randomized order." But different from this spirit, there was no concern to compare the two orders with a view to their differential effects on the course of health. Instead, addressed as "main outcome measures" were "cancer detection rate (yield), sensitivity, specificity, positive predictive value (PPV3) of biopsies performed and interval cancer rate." This content in the abstract is followed by a section on results, with a plethora of numbers in it.

One remarkable feature of those results was the peculiar conception of a diagnostic test's 'sensitivity' and 'specificity.' Ordinarily those two purported measures of a test's diagnostic 'accuracy' are taken to be conditional on the presence and absence, respectively, of the illness in question (Sect. 2.1). But this does not apply to a test's application in screening for breast cancer, as each instance of the screening cannot be classified according to the presence/absence of the cancer, but merely according to the cancer getting to be or not getting to be detected to be present. The authors substituted the latter for the former; and so they reported, for example, that the sensitivity for all three imaging tests combined is 100 % with a 95 % interval from 79 % to 100 %. But: a test's result inherently is positive in 100 % of the instances for which this positivity (rather than the cancer's presence) is a sine-qua-non criterion.

Another remarkable feature was the application of statistical tests of significance where they definitely do not belong. The context for testing statistical significance is the possibility that the 'null hypothesis' could in fact obtain, a priori and even conditionally on the result at hand. But when the result was, for example, that "Supplemental … ultrasound [testing] identified 3.7 cancers per 1000 screens," logical significance should have trumped statistical significance: a single case would have been sufficient to deny the null possibility. Attachment of "P < .001" to that result was thoughtless if not ignorant.

Ostensibly from those results was drawn the "Conclusion" that "The addition of screening ultrasound or MRI to mammography in women at increased risk of breast cancer resulted in not only a higher cancer detection yield but also in an increase in false-positive findings." This, however, was not a conclusion but a simple summary of the complex results. And the results in this meaning of them should have been confidently foreseen, thereby obviating this major 'clinical trial' with its wealth of major aberrations of the principles of screening-diagnostic research.

This study indeed was an original contribution but, sadly, not so much to genuine medical-scientific literature as to its highly prevalent ersatz, which Eric Topol terms not literature but litter-ature.

Study 2 – the other one of the two original studies for diagnosis appearing in the 2012 issues of *JAMA* – was:

> *Reference 2:* Min JK, Leipsic J, Pencina MJ et al (2012) Diagnostic accuracy of fractional flow reserve from anatomic CT angiography. J Am Med Assoc 308:1237–1245

According to its stated "Objective," this study was directed to diagnosis about "hemodynamically significant coronary stenosis," while under the study's "Design, setting, and methods" it was said to be about "ischemia," meaning myocardial ischemia. At issue thus was diagnosis about myocardial ischemia due to coronary stenosis, meaning the propensity to exertion-induced episodes of this, manifest in angina pectoris. In other words, the study was directed to the requisite knowledge-base of diagnosis about chronic coronary heart disease (Sect. 3.1).

In regard to chronic CHD, the diagnostic concern in this study was not about its presence/absence. It was, instead, diagnosis about whether a particular stenosis, identifiable by coronary angiography, is causal to the angina, in order that "percutaneous coronary intervention" – coronary angioplasty, that is – can be directed to such lesions alone.

Even more specifically, the concern was this lesion-specific diagnosis (about its role in angina) by noninvasive means – that is, without invasive coronary angiography augmented by measurement of the proportional pressure gradient across the stenosis, without this ICA-PPG.

The natural domain for this study (its object and then the study proper) is orientationally characterized by presence of chronic CHD (manifest in episodic angina pectoris) and at least one coronary stenosis as defined and identified by noninvasive, CT-based coronary angiography, CTCA.

For such a domain, the diagnostic probability function naturally would be designed to involve indicators based the particulars of the angina. Other indicators would address particulars of the stenosis, these in terms more comprehensive than merely the proportional extent of the artery lumen's narrowing (incl., e.g., the size of the artery). And it would involve indicators based on noninvasive stress tests. Besides, it now would involve the novel counterpart of the ICA-PPG derived from CT-based coronary angiography, the CTCA-PPG.

Essential in the study proper would be a series of cases from the designed domain; documentation of each of the domain-satisfying stenoses according to the designed PPF; and, of course, the generally most challenging element in diagnostic studies: determination of the truth about the presence/absence of the condition at issue, here about the role of the stenosis in the angina pectoris.

If an informative – suitably discriminating – DPF of this type could thus be developed, it would allow appropriate selection of the stenoses for angioplasty for the purpose of relief from the angina.

The authors, however, evidently had the idea that angioplasty (stenting) of specifically angina-causing stenosis likely would have the added benefit of preventing acute CHD and, thereby, deaths from CHD. They did not give the rationale for this less-than-obvious idea, surprising especially in the context of the authors noting that "recent randomized trials ... have identified no survival benefit for patients who undergo angiographically based coronary revascularization [refs.]." They did, however, refer to a trial which purportedly showed survival advantage from limiting angioplasty to stenoses characterized by ICA-associated PPG ≥ 0.20; but perhaps survival would be best if angioplasty be avoided in these instances as well. I'll discuss that trial as the first example in Sect. 7.5.

In line with the prevailing culture in diagnostic research, in the actual *study 2* the concern for suitably designed DPF was replaced by focus on a single test, the

CT-PPG; and serious determination of whether a stenosis is causal to the angina was replaced by the result of another single test, the ICA-PPG test giving a positive result in the meaning of PPG $\geq$ 0.20. The study was reduced to exploration of how good an ersatz the PPG from CTCA would be for its ICA counterpart, without discussing how diagnostic the latter is.

The study was "designed to evaluate the accuracy of [CTCA-PPG] to diagnose hemodynamically significant CHD, as defined by an invasive [PPG] reference standard, with a targeted population of patients with suspected native CHD who were referred for clinically indicated nonemergent ICA . . ." The clinical indications for ICA were left unspecified; and more notably, the case series ("population") from the referrals was not reduced to instances of actual CAD: the proportion of them with "angina within the past month" was only 77 %.

The PPG was measured by each of the two methods when the stenosis was in the range from 30 % to 90 %, while below and above this range the values were imputed to be 10 % and 50 %, respectively. Then, the results from both methods were dichotomized as either above or below 0.20. "In the per-patient analysis, vessels with the most adverse clinical status were selected to represent a given patient," ultimately in terms of that dichotomy.

In these terms, then, the core data were those in the attached Table 5.1, having to do with 252 patients. For it is in those terms that the investigators addressed their "Main outcome measures," specified this way: "The primary study outcome addressed whether [CTCA-PPG] plus CT could improve the per-patient diagnostic accuracy such that the lower boundary of the 1-sided 96% confidence interval of this estimate exceeded 70 %."

As is shown in that Table 5.1, the "accuracy" of the CTCA-PPG test was taken to be the proportion of the study instances in which the binary result of this test accords with that from the ICA-PPG test, which served as the "reference standard" – all of this strictly in line with the prevailing dominant culture of diagnostic research (Sect. 5.4). Reported as the core result was the non-invasive test's 73 % "accuracy" together with its two-sided 95 % imprecision ("confidence") interval (Table 5.1).

The main reason for the involvement of the 70 % reference point in the "main outcome measure" (above) may have been that "an array of prior studies . . . have demonstrated 70 % to be at the midpoint of reported diagnostic accuracies for stress imaging, depending on test type, patient population, and the disease prevalence [ref.]." The "Conclusion" was that "the study did not achieve its pre-specified outcome goal for the level of per-patient diagnostic accuracy," but the associated "Comment" was that the 73 % result together with its associated imprecision interval (Table 5.1) "establish a performance of [CTCH-PPG] that is within the range of conventional stress imaging testing."

Another reason likely was the felt need to carry out, and to report, the 'sample size determination' for the study. In the culture of this, as it is, the study size is not

Table 5.1 Core data from Study 2. Results of the dichotomized results (+, −) from study test and "reference standard" test in 252 patients, here denoted by Test 1 and Test 2, respectively

		Test 2		
		+	−	Total
Test 1	+	116	56	172
	−	13	67	80
	Total	129	123	252

Accuracy of Test 1: (116 + 67)/252 = 73 %; 95 % CI: 67 % − 78 %

designed to provide a given level of precision for the result – in this case on the study test's diagnostic "accuracy" – but a given level of statistical "power" to reject the 'null hypothesis' insofar as it in fact is false. To this end is needed a particular 'null hypothesis,' and this was taken to be that 70 % "accuracy" of the test under study. (This was explained not under "study design" but under "statistical analyses.") The report made no point about the relevance of this calculation as for the meaning of its core result (or any other results), as is usual in this aspect of the prevailing study-design culture.

The "Results" included also four other measures to characterize the diagnostic "accuracy" of the test under study, all of them based on the data given here in Table 5.1. They were the test's "sensitivity," 116/129 = 90 %; "specificity," 67/123 = 54 %; "positive predictive value, "116/172 = 67 %; and "negative predictive value," 67/80 = 84 %. This, too, is in accord with the prevailing dominant culture of diagnostic research.

Let us now think about all of this, specifically on the premise (false) that there is a justification for addressing a test's diagnostic informativeness when applied in isolation, and for doing this by comparing the test's dichotomized results with those of 'reference standard' test, the latter taken to be pathognomic about the presence/absence of the condition at issue in the diagnosis.

The fundamental question here is this ontologic one: Is it reasonable to think of a diagnostic test as having, in its solo application, a performance characterized by five parameters, constants across the subdomains of the test's domain of possible application?

In this vein, we need to ask for a start: Is the 'sensitivity' of the CTCA-PPG test for the angina-causing role of a coronary stenosis constant across the subdomains of the overall domain of the test's potential application and thus independent, most notably, of the particulars of the angina and the stenosis at issue? The answer in principle is that if there are subdomains such that the causality, when present, is relatively strong and others in which it is relatively weak, then there correspondingly are subdomains of relatively high and relatively low positivity rate for the test's result conditionally on the causality. Such distinctions can quite possibly be made at least on the basis of the severity, if not also the radiation pattern, of the angina and also according to the location of the stenosis at issue and the pattern of other stenoses.

The test's 'specificity' varies across subdomains if there are variations among them in the propensity of the test to have a negative result when the stenosis at issue is not causal to the angina. It is again not justifiable to presume that these variations do not exist.

The so-called diagnostic accuracy (not otherwise specified) of the test is a weighted average of its 'sensitivity' and 'specificity,' with weights proportional to the respective prevalences of the conditions in these (causality present and causality absent, respectively) in the domain at issue. Thus, whereas the results from the data in Table 5.1 for 'sensitivity' and 'specificity' are 0.90 and 0.54, respectively (cf. above), and the respective proportions for causality's presumed presence and absence are $129/252 = 0.51$ and 0.49, the result for the test's 'diagnostic accuracy' is $0.51(0.90) + 0.49(0.54) = 0.72$, which differs from the 0.73 in the table only on account of rounding-off errors.

Now, even if one (unjustifiably) takes the study test's 'sensitivity' and 'specificity' for the diagnosis of the causality to be constant over subtypes of the study subjects, one cannot take those weights to be constant from one study to another. Those $W_1 = 0.51$ and $W_0 = 0.49$ (above) reflect, for example, the fact that the prevalence of "angina within the past month" was only 77 %. Had the admissions into the study series reflected the idea that the diagnosis (with a view to possible 'revascularization') really deserves to be pursued only in those who have experienced angina within the past month, the weights might have been $W_1 = 0.60$ and $W_0 = 0.40$; and with these weights the result on the test's 'diagnostic accuracy' – presuming invariant 'sensitivity' and 'specificity' – would have been 0.60 $(0.90 + 0.40(0.54) = 0.76$.

Studies on a test's 'diagnostic accuracy' really need to involve a 'reference standard' for the positivity rate of the 'reference standard' test, reflecting a particular 'standard' domain of the study test's application and implying a particular pair of those weights, such as $W_1 = 0.60$ and $W_0 = 0.40$ in the example at issue here. In these terms the 'standard error' of that overall measure of the test's 'diagnostic accuracy' is

$$SE = \left[W_1^2 \hat{P}_1\left(1 - \hat{P}_1\right)/N_1 + W_0^2 \hat{P}_0\left(1 - \hat{P}_0\right)/N_0\right]^{1/2},$$

where $\hat{P}$ and $\hat{P}$ are the obtained empirical values of 'sensitivity' and 'specificity,' respectively, and N_1 and N_0 are the respective numbers of instances on which those empirical values are based.

If in the example at issue here the 'reference standard' for the testing's domain is characterized by $W_1 = 0.60$ and $W_0 = 0.40$, the result for the test's 'diagnostic accuracy' is 0.76 (above) with $SE = 0.032$, implying for the 95 % one-sided (*sic*) imprecision interval the lower bound of 0.71.

By the same token, the result for the test's 'positive predictive value' should not be derived from the data in Table 5.1 as the proportion $116/172 = 0.67$, but as 0.60 $(116/129)/[0.60(116/129) + 0.40(56/123)] = 0.75$. As for the 'negative predictive value,' analogously, $67/80 = 0.84$ should be replaced by $0.60(13/129)/[0.60$ $(13/129) + 0.40(67/123)] = 0.78$. In terms of the notation above, these two results are $W_1 \hat{P}_1 / W_1 \hat{P}_1 + W_0 \hat{P}_0$ and $W_0 (1 - \hat{P}_0) / [W_1 (1 - \hat{P}_1) + W_0 (1 - \hat{P}_0)]$, respectively.

These values are not predictive of anything; they are empirical values for the probability that the 'reference standard' test's result would have been positive/negative given the study test's positive/negative result, in the test's application in the domain characterized by those weights. Insofar as these probabilities are taken to be ones of causality, at issue is current existence of the causality, not future causality, much less inherently high future probability (on which any prediction would be based).

One more note on the 'statistical analysis' in this study. Among the "Patient and vessel characteristics" of the study subjects was given "Coronary calcium score, mean (SD), Agatson units" as "381.5 (401.0)." Anyone among the report's 20 authors should have edited this to 38 (40), and any statistician among them should have objected, vigorously, even to this edited version of the score distribution's description. For, mean together with 'standard deviation' defines only a distribution that is near-Gaussian, and when the SD is as large as the mean, the distribution is very, very skewed. Needed here is the generally applicable description in terms of centiles, say the 25th and the 75th in addition to the 50th (median).

And a final note on these test 'accuracy' studies at large. Seemingly without exception the study result for a given measure of this is unjustifiably written about in the style of saying, for example, that the test's 'sensitivity was' X%, instead of saying that the result for (the obtained 'estimate' of) it was X%. The mistake of believing in the existence of singular theoretical value for this must not be compounded by the mistake of writing as though each study result on it were identical to that value.

Chapter 6
Original Research for Scientific Etiognosis

Contents

6.0   Abstract

In medicine, 'etiology' no longer means the causal origin – etiogenesis – of sickness (overt) in its underlyng illness (somatic, hidden). It now mainly means the etiogenesis of illness; and correspondingly, etiognosis is doctors' esoteric knowing (probabilistic) about the etiogenesis of a case of illness that has occurred, specifically about whether an antecedent that was there was causal to the case of illness.

If the etiogenesis of an illness were a phenomenon (and hence observable), an etiognostic study would involve, simply, a series of cases of the illness in question, with histories on the patients in respect to the (potentially) etiogenetic factor at issue. But as causal connection is not observable, needed is a series of cases of the illness occurring in a defined experience of potential occurrence, coupled with a series of probes into the latter; and based on these two series, needed is addressing the rate of case occurrence in such a way that its relation to the antecedent in question can be viewed as a manifestation of the etiogenetic connection under study, in terms of causal rate ratio as a function of its modifiers.

O. S. Miettinen, *Toward Scientific Medicine*, DOI 10.1007/978-3-319-01671-9_6,   117
© Springer International Publishing Switzerland 2014

6.1 The Etiological Standpoint of Robert Koch

The philosopher K. C. Carter makes a strong statement about medicine in the very first paragraph of the Introduction to a medical book of his (ref. below):

> Of the numerous changes that have occurred in medical thinking over the last two centuries, none have been more consequential than the adoption of what Robert Koch called the etiological standpoint [ref.]. The etiological standpoint can be characterized as the belief that diseases are best controlled and understood by means of causes and, in particular, by causes that are *natural* . . . , *universal* . . . , and *necessary* . . . This way of conceiving disease has dominated medical thought for the last century. . . . [P. 1; italics in the original.]

Reference: Carter KC (2003) The rise of causal concepts of disease: case histories. Ashgate, Burlington

The prime example of this viewpoint in application is Koch's work on what is now known as *tuberculosis*, as he redefined it. "Koch found that adding potassium to the standard methylene-blue stain . . . revealed a new kind of bacillus in tuberculous materials. . . . Extensive inoculation tests supported the conclusion that the bacillus in question was the causal agent in tuberculosis. . . . Koch reclassified, as one disease, all the symptomatically and pathologically distinct cases in which the organism could be identified and excluded all other cases as not true tuberculosis [ref.] thereby, in effect, giving a new definition to 'tuberculosis' [ref.]." (Ref. above, pp. 133–135).

"Koch's monumental 1884 paper on the etiology of tuberculosis contains his most complete discussion of causation. . . . Here, and in his 1882 papers, Koch outlines a series of steps for proving causation. These steps can be summarized as follows:

Rt1. An alien structure must be exhibited in every case of the disease.

Rt2. The structure must be shown to be a living organism and must be distinguishable from all other organisms.

Rt3. The distribution of organisms must correlate with and explain disease phenomena.

Rt4. The organism must be cultivated outside diseased animals and isolated from all disease products that could be causally significant.

Rt5. The pure isolated organisms must be inoculated into test animals and the animals must then display the same symptoms as the original diseased animal." (Ref. above, pp. 135–136).

Consistent with the ideas advanced by Koch, my *Dorland's* (1992) defines *tuberculosis* as:

> any of the infectious diseases of man and animals caused by species of *Mycobacterium* and characterized by the formation of tubercles or caseous necrosis in the tissues.

The causal part of this definition is due to Koch, while before him tuberculosis was, simply, that "formation of tubercles or caseous necrosis in the tissues."

It was, specifically, for his work on tuberculosis that Koch was awarded the Nobel Prize for Physiology and Medicine, in 1905.

Koch's way of thinking about the causation of tuberculosis had notable precursors. For example, in 1835 "James Pagent discovered encapsulated worms in human muscle tissue that, by 1860, were recognized as the universal, necessary and sufficient cause of trichinosis" (ref. above, p. 25).

And his "etiological viewpoint" is with us today, well beyond the prevailing concept of tuberculosis (above). In my *Dorland's* the definition of *trichinosis* is:

> a disease due to infection with trichinae. It is produced by eating undercooked meat containing *Trichinella spiralis*. It is attended in the early stages by [a certain set of symptoms] and later by [another set of symptoms].

And the definition of *asbestosis* in this dictionary is:

> a form of lung disease (pneumoconiosis) caused by inhaling fibers of asbestos and marked by interstitial fibrosis of the lung varying in extent from minor involvement of the basal areas to extensive scarring; it is associated with pleural mesothelioma and bronchogenic carcinoma.

The definitions of tuberculosis, trichinosis, and asbestosis in *Dorland's* do, however, lack the necessary analogousness (beyond the "due to" and "produced by" duality in place of the "caused by" in the other two). By the conceptual and linguistic paradigms of trichinosis and asbestosis, 'tuberculosis' is a misnomer for *mycobacteriosis*. And for these three 'oses,' one possibility for analogous definitions in terms of *causation* is this:

– Mycobacteriosis: somatic anomaly (of, e.g., lungs) *caused by presence*, in the tissue, of mycobacteria;
– Trichinosis: somatic anomaly (of muscles) *caused by presence*, in the tissue, of trichinella worms; and
– Asbestosis: somatic anomaly (of lungs) *caused by presence*, in the tissue, of asbestos fibers.

Another causality-based possibility is this:

– Mycobacteriosis: somatic anomaly (of, e.g., lungs) *caused by exposure* to (airborne) mycobacteria;
– Trichinosis: somatic anomaly (of muscles) *caused by exposure* to (foodborne, live) trichinella worms; and
– Asbestosis: somatic anomaly (of lungs) *caused by exposure* to (airborne) asbestos fibers.

And a third possibility is this *acausal* set of definitions:

- Mycobacteriosis: somatic anomaly (of, e.g. lungs) *characterized* by presence, in the tissue, of mycobacteria (and, of course, the tissue changes caused by this presence);
- Trichinosis: somatic anomaly (of muscles) *characterized* by presence, in the tissue, of trichinella worms (and, of course, the tissue changes caused by this presence); and
- Asbestosis: somatic anomaly (of lungs) *characterized* by presence, in the tissue, of asbestos fibers (and, of course, the tissue changes caused by this presence).

That first one of these three sets of alternative definitions is, of course, in the spirit of Koch's "etiological viewpoint"; and it is to be noted that this viewpoint of focusing on causes that are universal, necessary, and sufficient actually was first, and very emphatically put forward by *Jacob Henle*, in 1844. Moreover, as Carter describes it (ref. above, pp. 24–25), this he did in the inaugural issue of *Zeitschrift für rationelle Medizin* (Journal of Rational Medicine), of which he was a co-founder.

It thus is fair, and indeed important, to consider this question: Which one of those three sets of definitions is *the set of choice, from the vantage of rationality*? That first set of 'definitions,' à la (Henle and) Koch, must be deemed rationally untenable, this on the ground that a cause of something inherently is extrinsic (and antecedent) to that something. In each of mycobacteriosis, trichinosis, and asbestosis the somatic presence of the agent is not causal to the somatic anomaly, as it is not extrinsic to the somatic anomaly; the agent's presence is an integral part – in the essence – of the somatic anomaly, intrinsic to it, in each of these three illnesses. As for tuberculosis/mycobacteriosis, as Carter points out, the agent got to be a universal, necessary, and sufficient cause of this illness simply because the illness was redefined (by Koch) so that this pattern of 'causality' got to be invariably the case.

That second set, too, is rationally untenable, for a different reason. A somatic anomaly 'defined' by its causal agent (necessarily extrinsic to it) actually is a somatic anomaly left undefined. In those three 'definitions,' nothing is specified about the essence – the always present, unique nature – of the anomaly. Those 'definitions' of illness are as empty of the necessary specificity (of somatic anomaly) as is 'definition' of 'tobacco disease' as any (*sic*) somatic anomaly that is caused by tobacco smoke.

Only that third, acausal set of definitions qualifies as a set of genuine definitions of illnesses, that is, of somatic anomalies with at least the potential of overt manifestation, in sickness (Sect. 5.1). Only they, thus, represent admissible conceptions of particular illnesses in the framework of truly rational medicine.

6.2 The Concepts of Etiology and Etiognosis

Before defining the concept of etiology, let us develop relevant background for this by exploring the concepts of genesis, carcinogenesis, oncogenesis, and pathogenesis. My *Dorland's* (1992) says that genesis is: "the coming into being of anything: the process of originating." For carcinogenesis its definition is: "the production of carcinoma," and for oncogenesis it is: "the production or causation of tumors." *Pathogenesis* it defines as:

> the development of morbid conditions or of disease; more specifically, the cellular events and reactions and other pathologic mechanisms occurring in the development of disease.

As for *etiology*, then, this *Dorland's* specifies it thus:

> the study or theory of the factors that cause disease and the method of their introduction into the host; the causes or origin of a disease or disorder. Cf. *pathogenesis*.

Another definition for etiology also is of note. Etiology is of central concern in epidemiological research, and the International Epidemiological Society defines it thus (ref. below):

> Literally, the science of causes; in common usage, cause.

Reference: Porta M (Editor), Greenland S, Last JM (Associate Editors) (2008) A dictionary of epidemiology. A handbook sponsored by the I.E.A., 5th edn. Oxford University Press, Oxford

This set of definitions is unsatisfactory even if examined only uncritically; for, inconsistencies abound.

As for *genesis*, those definitions of carcinogenesis and oncogenesis are inconsistent not only with each other but also with what the same dictionary defines to be the essence of genesis per se. There are experimental ways of "production" of carcinomas and other tumors; but these are not examples of carcinogenesis or oncogenesis, of the genesis of these anomalies. Genesis of cancer, say, is its "coming into being," but not inherently and solely the "process" of its originating.

Histologically, carcinogenesis is the process of cellular changes that start from normal tissue and end with malignant neoplasia; and genetically the counterpart of this is the process of the accumulation of mutations with the end result of proto-oncogene having been activated into oncogene and suppressor genes having been inactivated. But there also is, very importantly, the causal aspect to carcinogenesis. The former alone is "the process of originating" of cancer, the latter being the driver of this process.

So, more generally, while the genesis of an illness is the coming into being of the illness, this has two dimensions: the *descriptive* dimension of how, and the *causal* dimension of why. The former is "the process" – "the pathologic mechanism" – of the genesis of the illness, a matter of successive pathological phenomena, while the causation of this is not phenomenal (but noumenal). Carcinogenesis is the genesis of cancer in this dual meaning of 'genesis.' (In the Bible, one of the two stories of Genesis – the Elohist rather than the Yahwist one – describes the process of successive stages of the world's coming into being, and it ascribes these to their cause, namely Elohim, or God.)

And to be understood about genesis, both in general and in these particulars, is that it is a *retrospective* concept: genesis of an illness presupposes that the illness actually has come into being, and the genesis of this outcome is the how and why of this – backward from the time of this outcome. (Cf. Genesis in the Bible.)

That duality in the genesis of an illness is linguistically expressed by the *pathogenesis* versus *etiogenesis* duality. Thus, pathogenesis of an illness is the descriptive aspect of its genesis, the way in which it developed (in a particular case) or develops (in general), while etiogenesis of an illness is the causal aspect of its genesis, the causal explanation of its pathogenesis. (The term 'etiogenesis' is a rather recent neologism I've introduced.)

Etiology of an illness is, simply, the causal origin, the etiogenesis, of it; the term thus is a synonym of 'etiogenesis.' But the term 'etiology' is rather problematic in that synonymical usage. 'Etiology' rhymes with 'pathology,' while only 'etiogenesis' rhymes with 'pathogenesis,' the concept being intimately related to that of pathogenesis (cf. above). And only the term 'etiology' sustains such misapprehensions about its referent concept as the ones above: that it is "the study or theory of the factors that cause disease," or that it is "literally, the science of causes." There is no such singular "study or theory," and there is no such singular "science." There are etiogenetic challenges peculiar to oncology, others peculiar to cardiology, etc.; and etiology is no more literally the science of causes than, say, tautology is literally the science of unnecessary repetition.

Etiognosis, then, is a doctor's knowing (esoteric) about a particular etiogenesis of a case of illness, just as diagnosis is such knowing about the presence of a particular type of illness. It naturally focuses on particular, potentially etiogenetic antecedents of the case, just as diagnosis focuses on illnesses that represent potential explanations of the patient's sickness. And in respect to any given potentially etiogenetic antecedent of a case of illness, one that actually was there, etiognosis is knowing about the probability (objective) that it actually was causal to the illness, just as diagnosis is knowing about the probability that the illness in question actually is present. (Like 'etiogenesis,' 'etiognosis' is a neologism I've adduced.)

This *concept of etiognostic causality* is quite different from that in the 'etiological standpoint' of Robert Koch, addressed in Sect. 6.1 above. An antecedent was causal to a case of the illness that has occurred if the illness would not have occurred (at the time it did occur) but for the antecedent having been there, ceteris paribus; in particular, it would not have occurred if the alternative of the antecedent in the causal contrast had been present, ceteris paribus. The concept almost never has to do with a cause that is necessary and sufficient (cf. Sect. 6.1 above). Instead, the cause typically was neither necessary nor sufficient. It completed a sufficient cause while its alternative would not have.

The discussion above has been about etiogenesis of and etiognosis about an illness, with no reference to *sickness*. But actually, a clinician's etiognosis typically is about possible etiogenesis of the patient's sickness in the medications the patient has been using, etiognosis about such sickness rather than some illness. This is why non-illness etiogenesis is prone to be involved in the differential-diagnostic set (Sect. 1.2). In point of fact, though, even diagnosis about an illness can be said to be etiognosis about sickness when at issue is presentation with sickness.

6.3 Hypothetical Simple Etiognostic Study

It is instructive to approach the topic of etiognostic research initially in the context of *imagining that causation is a phenomenon* (rather than being only a noumenon, not an empirical concept but a conception a priori, as Kant described it). On this counterfactual premise, any given instance of antecedent-subsequent conjunction can be observed as representing or not representing a causal conjunction.

On this premise, an etiognostic study would involve only a series of cases of the outcome (illness or sickness) in question; and in this *case series* the focus would be on the subset of the cases in which, associated with the case, is a positive history for the potentially etiogenetic antecedent in question. The proportion of cases occurring in association with the antecedent would not generally be seen to be of scientific interest, as it commonly would be understood to reflect the particularistic (spatio-temporally specific) frequency of the antecedent's occurrence. The use of thalidomide is a case in point: this medication for 'morning sickness' (in pregnancy) was withdrawn from the markets once its singular role in the etiogenesis of a major birth defect (phocomelia) was discovered. And even in general, no medication use has a frequency that is anywhere near constant over place and time, not even subsequent to its introduction and use conditionally on its 'official' indications.

The cases occurring in association with the antecedent in question naturally would be classified according to whether the antecedent was causal to the case coming into being, and the research interest would focus on the proportion of the outcome-antecedent conjunctions that actually are causal. The interest would focus on this *factor-conditional etiogenetic proportion.*

This FEP is prone to vary by various particulars of both the antecedent and of the 'host' of the outcome in question, by various modifiers of this proportion's magnitude. Thus the generic object of this hypothetical etiognostic study, and hence of any actual etiogosstic study, naturally this FEP *as a joint function of select modifiers* of its magnitude.

6.4 The Elements in Actual Etiognostic Studies

Just as in the hypothetical situation addressed in Sect. 6.3 above, a *case series* of the illness or sickness at issue is central to any actual etiognostic study as well; and just as in that hypothetical study, the case series in an actual etiognostic study provides information about the factor-conditional etiogenetic proportion in respect to the etiogenesis at issue. The challenges in this arise from the need to work with actual phenomena manifesting that FEP (ultimately as a function of its modifiers).

In an actual etiognostic study the case series needs to be used as the source of numerator inputs into the outcome's *rates* of occurrence in a particular experience, in a particular aggregate of population-time, or series of person-moments, constituting the *study base*, separate rates for two segments of the study base: the *index* rate for the index segment of the study base, characterized by positive history for the factor in question, and the *reference* rate for the reference segment of the study base, characterized by positive history for the alternative to the factor in question (cf. Sect. 6.2). This leaves out a third, the 'other,' segment of the study base.

Denoting these empirical index and reference rates by R_1 and R_0, respectively, the result of interest is, to a first approximation, the 'crude' one of

$$\widehat{FEP} = (R_1 - R_0)/R_1$$
$$= [(R_1/R_0) - 1]/(R_1/R_0)$$
$$= \left(\widehat{RR} - 1\right)/\widehat{RR},$$

where $\widehat{RR}$ denotes the rate ratio of interest. This is the result of interest on the generally untenable premise that the index and reference segments of the study base have closely similar, 'balanced' distributions according to potential *confounders* – extraneous determinants of the outcome's rate of occurrence conditionally on the reference category of the etiogenetic determinant at issue.

In deriving that crude empirical rate-ratio ($\widehat{RR}$) it is not necessary to determine the actual sizes of the index and reference segments of the study base; nor is this determination generally feasible, given the infinite number of person-moments constituting the population-time of the study base. It is enough, and necessary, to obtain numbers proportional to these, as inputs to the corresponding quasi-rates: $Q_1 = C_1/B_1$ and $Q_0 = C_0/B_0$ based on the case series of size $C = C_1 + C_0 + C'$ coupled with a *base series* – a fair sample of the person-moments constituting the study base – of size $B = B_1 + B_0 + B'$. For on the basis of these two series,

$$
\begin{aligned}
\widehat{RR} &= (C_1/B_1)/(C_0/B_0) \\
&= (C_1/C_0)/(B_1/B_0) \\
&= C_1/(B_1 C_0/B_0) \\
&= O/\hat{E},
\end{aligned}
$$

where $O = C_1$ is the '*observed*' number of cases associated with the index history, and $\hat{E} = B_1 C_0/B_0$ is the empirical value for the corresponding '*expected*' number, which is conditional on (the counterfactual of) the reference rate having characterized the index experience.

As the ultimate $\widehat{FEP}$ is determined by the unconfounded counterpart of this $\widehat{RR}$, a core challenge in etiognostic studies is the derivation of a suitably *adjusted* counterpart of this crude $\widehat{RR}$, based on suitable adjustment of the $\hat{E}$ in it.

6.5   Confounder Adjustment of Rate Ratio

A study result's adjustment for particular confounders – making it unconfounded by select confounders of the study base – is termed *control* of confounding (distinct from prevention of confounding, of confounding becoming a feature of the study base). The core *principle* in this is the following: An extraneous determinant of the outcome's rate of occurrence (in the reference experience), however it may confound the study base, does not confound the study result – the empirical value of a comparative measure, notably the empirical $\widehat{RR}$ – when that result reflects solely the outcome-antecedent association conditional on (the various realizations of) that confounder of the study base.

Thus, if the study data are cross-stratified by the set of confounders to be controlled, and if for the jth stratum of this cross-stratification the data are $C_{1j} = O_j$ and $\hat{E}_j = B_{1j} C_{0j}/B_{0j}$ (cf. Sect. 6.4 above), then the $\widehat{RR}$ adjusted for these confounders is

$$\widehat{RR}^{\,*} = \sum_j C_{1j} / \sum_j \left(B_{1j} C_{0j} / B_{0j} \right).$$

The asterisk in this denotes not only that the measure is adjusted for certain confounders but, more specifically, that the adjustment is a matter of taking the total 'observed' number of index cases ($O = \sum_j C_{1j}$) as a given and adjusting the 'expected' number by suitable conditioning.

This particular adjustment is in accord with what is needed for study of the factor-conditional etiogenetic proportion introduced in Sect. 6.3. It differs, most notably, from the Mantel-Haenszel measure introduced into epidemiological research in 1959:

$$\widehat{RR} = \sum_j \left(C_{1j} B_{0j} / T_j \right) / \sum_j \left(C_{0j} B_{1j} / T_j \right),$$

where $T_j = C_{1j} + B_{1j} + C_{0j} + B_{0j}$.

That $\widehat{RR}^{\,*}$ is, in epidemiological terminology, the *standardized* $\widehat{RR}$ in the particular meaning that the weights of the standardization (of the quasi-rates, Qs; Sect. 6.4 above) have been derived from the index experience, as $W_j = B_{1j}/\sum_j B_{1j}$:

$$\begin{aligned}
\widehat{RR} * &= \sum_j C_{1j} / \sum_j \left(B_{1j} C_{0j} / B_{0j} \right) \\
&= \sum_j B_{1j} \left(C_{1j} / B_{1j} \right) / \sum_j B_{1j} \left(C_{0j} / B_{0j} \right) \\
&= \sum_j B_{1j} Q_{1j} / \sum_j B_{1j} / Q_{0j} \\
&= \sum_j \left(B_{1j} / \sum_j B_{1j} \right) Q_{1j} / \sum_j \left(B_{1j} / \sum_j B_{1j} \right) Q_{0j} \\
&= \sum_j W_j Q_{1j} / \sum_j W_j Q_{0j} \\
&= Q_1 / Q_0^*.
\end{aligned}$$

A point of note about this $\widehat{RR}^{\,*}$ also is the relation of each $\hat{E}_j$ in it to its corresponding stratum-specific $\widehat{RR}$:

$$\begin{aligned}
\hat{E}_j &= B_{1j} C_{0j} / B_{0j} \\
&= C_{1j} \left(B_{1j} C_{0j} / B_{0j} C_{1j} \right) \\
&= C_{1j} / \left[\left(C_{1j} / B_{1j} \right) / \left(C_{0j} / B_{0j} \right) \right] \\
&= C_{1j} / \widehat{RR}_j.
\end{aligned}$$

This means that the contributions not only to the $O = C_1$ set but also to the corresponding $\hat{E}$ derive only from strata with $C_{1j} > 0$, as is implied by the hypothetical situation addressed in Sect. 6.3. (This is not true of the Mantel-Haenszel $\widehat{RR}$, above.)

An added implication is that the contributions to the $\hat{E}$ from the relevant strata (for which $C_{1j} > 0$) tend to be highly variable by chance when the stratum-specific data are very sparse: $\hat{E}_j = 0$ if either $B_{1j} = 0$ or $C_{0j} = 0$ while $B_{0j} > 0$; $\hat{E}_j = \infty$ if $B_{0j} = 0$, $B_{1j} > 0$, and $C_{0j} > 0$; and $\hat{E}_j$ is undefined if $C_{0j} = B_{0j} = 0$, which makes $\widehat{RR}^*$ undefined.

Yet another implication is the principal one, namely this:

$$\widehat{RR}^* = \sum_j C_{1j} / \sum_j \left(C_{1j} / \widehat{RR}_j \right)$$
$$= C_{1j} / \sum_j \left(1 / \widehat{RR}_j \right),$$

where the *summation is over the index cases* (C_1 in number) and $\widehat{RR}_j$ is the empirical rate ratio specific to the confounder profile (individual 'stratum') of the jth index case, suitably derived. This opens the possibility to avoid the inefficiency/imprecision of the cross-stratification approach, by deriving those $\widehat{RR}_j$ values by means of *modeling* – designing a model for the outcome's rate of occurrence, fitting this to the study data (on the case and base series), and deriving the $\widehat{RR}_j$ values (and hence the $\widehat{RR}^*$) from this.

The *direct objects* of an etiognostic study thus get to be the relevant parameters in the designed model for the outcome's rate of occurrence in the designed domain of this.

6.6 Etiognostic Studies' Objects Design

Etiognostic research for clinical medicine would naturally focus on the development of the knowledge-base of *pharmacoetiognosis* – knowing about the etiogenesis of an 'adverse event' in a patient in respect to a particular medication use by the patient. For when a patient has experienced an AE such as acute agranulocytosis or incipient jaundice, the doctor needs to know about the probability of causal role in this for each of the medications the patient has been using, 'over-the-counter' medications included. Pharmacoetiognostic research for a particular discipline of medicine would naturally be further focused on such AEs as are inherently of concern in this particular discipline of medicine (à la agranylocytosis as a concern in hematology and incipient jaundice as a concern in hepatology), and on such medications as are commonly prescribed in this discipline (à la anti-inflammatory medications in rheumatology). And beyond AEs, pharmacoetiognostic concerns include adverse states of health possibly initiated and sustained by medication use, aberrant mental states, for example.

When the concern is with the etiogenesis of an *event*-type entity (illness or sickness), a model is designed (as to its form, distinct from the empirical content of that form, to be derived by the study) for its rate of occurrence, in a designed domain, generally as a matter of this event's *incidence density* of occurrence: number of cases per unit amount of population-time (1,000 person-years, say). In general, the appropriate model is *log-linear*:

$$\text{Log}(\text{ID}') = B_0 + \sum_i B_i X_i,$$

where ID' is the numerical element in the rate (leaving out the time-dimensioned unit of it, such as $[10^3 y]^{-1}$), and the independent variates have to do with (i.e., are based on) the etiogenetic histories of interest, confounders of the causal rate-ratio, and modifiers of the causal rate-ratio (its magnitude).

In respect to the medication use at issue, the causal role, if any, in an AE commonly is the precipitation of an acute idiosyncratic reaction to molecules of the medication present in the tissues at the time. Therefore, the causal history generally is about very recent use of the medication, a matter of days only; and with this temporal referent the relevant history generally is of the binary, all-or-none type rather than quantitative (with a view to dose–response).

So, commonly,

X_1: indicator of *recent* (as defined) use of the medication (1 if used, 0 otherwise).

The propensity for an idiosyncratic reaction generally manifests itself relatively early in the medication's use by the susceptible person; and once the reaction does occur, the medication's use, if recognized as causal, is discontinued and subsequently never reintroduced. Thus, in general, the longer the use of the medication, the more assured is the person's ability to tolerate it, the more completely the susceptibles have been depleted from among the continuing users. This means the need to define (in the simplest possible terms)

X_2: the aggregate duration of the medication's *previous* uses (the numerical part of this; number of years, say), and

X_3: $X_1 X_2$,

so that *modification* of the effect of the medication's recent use (in the model's domain) gets to be represented by $B_3 \neq 0$ in the simple case of no other modifiers being allowed for and

X_4, etc.: non-modifier confounder variates,

with no product terms based on these together with X_1.

This simple model implies the log difference between the index and reference rates (ID'_1, for which $X_1 = 1$, and ID'_0, for which $X_1 = 0$) to be $B_1 + B_3X_3$, so that the incidence-density ratio is the exponential of this difference:

$$IDR = \exp(B_1 + B_3X_3).$$

This implication, in turn, illustrates *the power of modeling*: the pair of empirical values for those two parameters provides for specification of the $\widehat{IDR}_j$ values corresponding to each of the different X_3 values associated with the C_1 index cases from the study, whatever will be their number, thus providing for enhanced-precision quantification of $\widehat{RR}^*$ (cf. Sect. 6.5 above).

While the invocation of modeling provides for avoidance of the inefficiency of confounder control in the framework of cross-stratification of the data by the confounders (Sect. 6.4), thus providing for control of more confounders and each quantitative one of them more closely, the sought-for superiority of it is genuinely achieved only insofar as the designed model is *realistic* in its form (for the referent domain of it). For example, while in the simple model above the aggregate of the previous uses is addressed in terms of the total duration of this, perhaps the cumulative amount (rather than mere duration) of the medication's previous use is a more meaningful determinant of the extent to which previous use of the mediation signifies freedom from propensity to have an adverse reaction to its current use; and there generally is a need to provide (realistically) also for modifies other than the medication's previous use.

6.7 Etiognostic Studies' Methods Design

An etiognostic study's methods design is, in general, bi-phasic. The first phase is the development of the *operational counterpart of the designed model* for the outcome's rate of occurrence (in the model's referent domain), while the second phase is design of the study proper – the acquisition of documented experience of the operationalized model's form and, based on this, acquisition of empirical values for the object parameters in the model (distinct from the parameters whose quantification merely serves the control of confounding).

Among the methodologic problems that may need to be solved in the operationalization of the model designed without regard for practicalities are the implications of 'medicalization' (Illich): relatively heavy consumers of healthcare tend to have more than the usual frequency of recent use of any particular medication, and they also tend to have more than the proportional representation among those with a case of an adverse event (or state) of health coming to medical attention, even when this outcome never is medication-induced. This means that a positive history for the use of a particular medication tends to be more common in an etiognostic study's case series than in its base series even when the medication's use is not etiogenetic to the outcome. But: this tendency presumably vanishes if the outcome for the operationalized model is defined as a *severe-and-typical* subtype of the entity in general, since a case of the thus-defined outcome presumably comes to medical attention and gets to be diagnosed regardless of the person's degree of medicalization. (That the degree of medicalization bears on the histories of the medication's use is not a validity concern.)

That restriction in the definition of the outcome entity also serves validity assurance in a more general and more fundamental regard, namely as to the *basic architecture* of the study, constituted by the case-and-base-series duality in it. The essential requirement for validity in this regard is that these two series be a coherent pair in the sense of having the same – a shared – referent, the study base in this role. This generally requires complete identification of the cases (as defined) in a defined study base together with fair sampling of the latter. When the source base (containing the study base) is given a direct definition, its fair sampling tends not to be problematic; but complete identification of the cases occurring in it is greatly facilitated – if not made possible only – by focus on cases that are severe-and-typical. And when the source population is defined indirectly – as the catchment population of the scheme of case identification – with this definition of the cases it takes a more concrete form and is thereby subject to more valid sampling, while the case series is complete by definition.

The validity challenge potentially arising from medicalization in pharmacoe-tiogenetic studies may look rather like one of confounding. Medicalization, however, is not a determinant of the (rate of) occurrence of the actual outcome (illness) at issue, but only of the occurrence-conditinal propensity of it being detected. Thus, at issue is not confounding bias but documentation bias. (Selection bias, concerning the study base as such, is yet another matter.)

Another potential source of documentation bias is *etiogenetic suspicion* induced by the factor under study, when this suspicion bears on the diagnosis (à la oral contraceptive use in the diagnosis about pulmonary embolism). And the solution to this problem too is its prevention by that restricted definition of the outcome entity, rather than quantification and control of the degree of suspicion.

When an actual confounder is quantitative but not subject to adequate control because of infeasibility of meaningful quantification of it, its quantity may have a well-quantifiable zero level of reasonably common occurrence. An eminent example of this in pharmacoetiognostic research generally is severity of contraindications. Confounding by something of this genre generally is preventable by restriction of the study objects' domain to the zero level of the confounder.

Confounding fundamentally being a feature of the index-reference contrast in the study base, it needs to be either documented-and-controlled or prevented in order to achieve an unconfounded result; but if neither one of these is feasible, it may be reasonable to *change the contrast* in the operationalization of the designed model and thereby to achieve unconfoundedness of the study base by that extraneous determinant of the outcome's rate of occurrence. Study of Reye's syndrome's possible etiogenesis in aspirin use in the context of febrile illness in childhood may be an example of this. This hypothesis was tested in several high-profile studies, and in these there was concern about confounding by the level of fever, but documentation of this level was very problematic because of the antipyretic effect of aspirin use. Perhaps the use of aspirin should have been contrasted with the use of acetaminophen, for example, as study of this contrast would have been free of confounding by level of fever (the indications for the contrasted medication uses being the same as for the level of fever).

As for the rest of the methods design of pharmacoetiognostic studies, a suitable introduction may be constituted by review of the status quo in 'pharmacoepidemiologic' study in the section below.

6.8   Some Recent Studies Critically Examined

Etiognosis being principally a concern in respect to possible iatrogenesis of a case of illness or sickness, most commonly by far from the use of a prescribed medication, I sought example studies specifically of the pharmacoetiognostic type, again from the 2012 issues of *JAMA*. The first and only one of these was this:

> *Reference:* Etminan M, Forooghian F, Brophy JM et al (2012) Oral fluoroquinolones and the risk of retinal detachment. J Am Med Assoc 307:1414–1419

The "Objective" was said to have been "To examine the association between the use of fluoroquinolones and the risk of developing a retinal detachment." This suggests that the objective was to study (*sic*) the effect (*sic*) of those medications' use on (*sic*) the risk of developing that adverse event; that this was a prognostic – intervention-prognostic – study.

But the "Design, setting, and patients" of the study were said to have been these: "Nested case–control study of a cohort of patients, who had visited an ophthalmologist … Retinal detachment cases were defined as … Ten controls were selected for each case …" Clearly, thus, at issue actually was an etiognostic study, one whose generic object was the etiogenesis of that adverse event in its antecedent use of those medications.

The "Main outcome measure" in this study was said to have been "The association between [the adverse event and its antecedent use of the medications]. Actually, the outcome – in the epidemiological meaning of the term – was that adverse event; and the "measure" of this was "a procedure code for retinal repair surgery . . ."

This study was said to have been a "case–control study" (cf. above), this characterization of it presumably intended to distinguish it from a 'cohort study' of etiogenesis; but this traditional and still-commonly perceived duality is a fallacy. Any proper study of the etiogenesis of an adverse event of health is anchored to a study base – an aggregate of population-time constituted by a dynamic (open-to-exits) population's course over a span of time; and it involves identification of the events in question occurring in this study base as well as coupling the resulting case series with a suitable base series: a fair sample of this study base, the latter constituting the referent of the result of the study (Sect. 6.6).

When said was that "From a cohort of 989 591 patients, 438 cases of retinal detachment and 43 840 controls were identified," the meaning was this: The study base was imbedded in a source base of population-time constituted by the course of a cohort (closed-for-exit population) over time; and from the study base within this source base, the case series was identified and a base series – not a group of "controls"! – was selected (not "identified").

The study was 'nested' in the study base, which in turn was 'nested' in the source base. When the point was made that this etiogenetic study was of the "nested" variety of them, implied was that this was a notable feature of it, implying its absence from some other admissible "case–control" studies. But a proper etiognostic (or other etiogenetic) study inescapably involves a study base (Sect. 6.4). In this study the source population had a direct, primary definition; but the alternative to this is the source population's indirect, secondary definition – as the catchment population implied by the direct/primary definition of the way in which the case series will be assembled (Sect. 6.7 above).

It was strikingly awkward to refer to the directly-defined, cohort-type (rather than dynamic) source population of this etiognostic study as "The nested case–control cohort." For it was, simply, the (cohort-type) source population for the etiognostic study (which involved the case series and a base series from the study base imbedded in the source base formed by the population-time of this cohort's follow-up).

Said was that "Controls were selected using density-based sampling, an approach that allows for the close approximation of the odds ratio to the rate ratio [ref.]." But there was no special, "density-based" sampling for the base series in order to be able to obtain a close approximation to the rate ratio, nor was there a need for some special type of sampling to this end. The rate ratio at issue here was the incidence-density ratio $IDR = \mathrm{ID}_1/\mathrm{ID}_0$ in the study base, where ID_1 is the index rate $C_1/B_1{}^*$ and ID_0 is the reference rate $C_0/B_0{}^*$, each representing the number of cases divided by amount of population-time. The base series gives B_1 and B_0 as tallies stochastically proportional to $B_1{}^*$ and $B_0{}^*$, respectively. For the $\widehat{IDR}$ documented thus is the value $(C_1/B_1)/(C_0/B_0)$ – with no "approximation of the odds ratio to the rate ratio" (Sect. 6.6).

With these fundamentals not understood by the investigators, they miscalculated the crude (unadjusted) $\widehat{IDR}$s. Thus, based on $C_1 = 145$, $B_1 = 3939$, $C_0 = 275$, and $B_0 = 40787$, they did not calculate the ratio of the two quasi-rates (C_1/B_1 and C_0/B_0) and thus obtain $\widehat{IDR} = 5.46$. Instead, they calculated $[(C_1/(C_1 + B_1)]/[C_0/(C_0 + B_0)]$, obtaining 5.55.

But not only was that calculation wrong; so too was the contrast it addressed. Contrasted was current use of the medications with no use of them within the past year, when the correct contrast would have been 'current' (meaning recent) use versus no 'current' use.

For the IDR having to do with the 'current'-use contrast, earlier use would have been important to consider as a (likely) modifier of that rate-ratio's magnitude (Sect. 6.6). But, very notably, no potential modification of the IDR's magnitude was considered.

Apart from the rate ratios for different contrasts regarding histories of the medications use, the investigators addressed "The absolute increase in the risk" of the outcome event. As for the meaning of this, said under "Statistical analysis" was, simply, that this "equaled the estimated incidence in the users (RR × incidence among nonusers)." And with the calculation nowhere explicated or shown, said under "Results" was, simply, that "The absolute increase in the risk of retinal detachment was 4 per 10 000 person-years (number needed to harm = 2 500 computed for any use of fluoroquinolones)."

Pertaining to this mystery, said under "Comment" was this:

The incidence of retinal detachment is estimated to be 12 per 100 000 patients [sic] annually in the United States [ref.]. Given an approximate exposure prevalence of 10%, and assuming a similar risk increase in the general population, the population attributable risk would be estimated to be approximately 4% [sic]. [Cf. "4 per 10 000 person-years" above.]

Very curiously, that overall ID was in reference to the United States as a whole rather than the population-time of the source base or the study base, which the report nowhere specified or even implied. And why use some unspecified "approximate exposure prevalence of 10%" rather than the rate documented in the study itself?

My best guess is that the reference ID actually was calculated as the solution (for X) of the equation below, involving that national ID together with the exposure prevalence (0.6 %) and $\widehat{IDR}$ (4.5) from the study itself:

$$0.006(4.5)X + (1 - 0.006)X = 12/10^5 y.$$

This implies $ID_1 = 53/10^5 y$ and $ID_0 = 12/10^5 y$ and, thus, the approximate rate difference of $40/10^5 y = 4/10^4 y$.

This reported result is not a difference between two "risks" ("absolute" or "attributable") but explicitly between two empirical IDs. (Risk is a theoretical rather than empirical quantity – the probability of an adverse event over a defined period of prospective time. For a prospectival time from $T = 0$ to $T = t$, it is determined by the theoretical ID_t this way: $1 - \exp\left[-\int_0^t ID_t dt\right]$.)

Looking back at all of this, the authors said under the "Comment" that "Our study has several strengths." The first one of these purportedly was that "we had a homogenous population of nearly 1 million patients ..." But relevant to the "strength" of the results is not the source population but the study population and, more specifically, the study base. The meaning and relevance of the source population having been "homogenous" were left unspecified; and as for size, relevant for the study results' precision was not the size of the source cohort nor even the size of the actual study base, but the number of cases from the study base with the index history. (Cf. $\widehat{IDR}$ calculation above.)

Another "strength" was said to be likely freedom from "confounding by indication," as "this is unlikely in this study because the factors related to the prescribing of oral fluoroquinolones are not usually [sic] known to be associated with retinal detachment [ref.]." But: what about confounding – and modification too – by the medication's use prior to the most recent 7 days? The more there was of this earlier use, the clearer is the patient's ability to tolerate the medication without suffering the retinal detachment from it; and the recent users and non-users of the medications cannot be presumed to have similar histories of earlier use of the medications in question.

The authors also made the countervailing point that "This study has several limitations," proceeding to specify some of them. But they naturally did not refer to the many misunderstandings and linguistic as well as other misrepresentations I've sketched above, nor did they make a point of what I regard as the principal limitation: failure to design and then address a rate (ID) function with suitable representation of the history contrast(s), conditionality on potential confounders, and potential modification of the magnitude of the (confounder-conditional) rate ratio (IDR) across the history contrast(s).

This example study is amply illustrative of how the prevailing theoretical understandings of etiognostic research are not much better developed among etiognostic researchers than are those of diagnostic research among diagnostic researchers (Sect. 5.10) – despite the central role of etiogenetic studies in population-level (rather than laboratory-based) epidemiological research.

Chapter 7
Original Research for Scientific Prognosis

Contents

7.0   Abstract

The concept of prognosis, like that of diagnosis, remains poorly developed in medical thought and writings, while even the term 'etiognosis' remains a neologism. In line with this, the theory of prognostic practice remains poorly developed, and the theory of research for scientific prognosis remains underdeveloped as a consequence of this. Emblematic of the lack of guidance from well-developed theory is the common focus in experimental studies on intervention-prognosis on 'hazard ratios' rather than intervention-conditional prognostic probability functions, which scientific prognosis would depend on.

Upon common misapprehensions of the concept of prognosis having been cleared away, the general essence of a prognostic study is implied by the genuine concept of prognosis. As with diagnostic and etiognostic studies, the essential features of a prognostic study are to be taken as a-prioristic givens, rather than matters of study design, and the challenges are ones of objects design and methods design beyond the necessary givens.

Methodologically, the key challenge has its resolution in bringing understanding of the essence of the etiognostic study to bear on that of the prognostic study.

O. S. Miettinen, *Toward Scientific Medicine*, DOI 10.1007/978-3-319-01671-9_7,
© Springer International Publishing Switzerland 2014

7.1 Misapprehensions about Prognosis

According to my *Dorland's* (1992), the denotation of '*prognosis*' is:

> a forecast as to the probable outcome of an attack of disease; the prospect as to recovery
> from a disease as indicated by the nature and symptoms of the case.

This follows the etymologic note of "[Gr. *progōnsis* foreknowledge]." So, while
etymologically at issue is a type of knowledge, this medical dictionary presents
prognosis as being a type of forecast for one and a type of prospect for another. This
echoes the dual concept of diagnosis in this dictionary: a type of determination for
one and a type of art for another, with its associated etymologic note referring to
knowledge as the meaning of the Greek *gnōsis* (Sect. 5.1).

Now, that "authoritative" medical dictionary is as *mistaken* about the concept of
prognosis as it is about that of diagnosis, manifesting misapprehensions about these
central concepts of medicine even in the most proximal terms. A doctor (L. *doctor*,
teacher) teaches the client about their health on the basis of the *esoteric knowing* (s)
he has attained about it – of gnosis, that is (Sect. 1.1). The teaching naturally
presupposes knowing about the client's health first as to what is, then perhaps about
why it is, and finally about what will be – its predicates thus being diagnosis,
etiognosis, and prognosis, respectively. Pursuit of these three types of esoteric
knowing is, incontrovertibly, in the essence of being a doctor; the respective
three terms are eminently apposite for them; and no alternative English-language
terms to express them are offered in my *Dorland's* or anywhere else.

It thus must be understood, for most proximal orientation, that prognosis is a
doctor's *esoteric knowing* about the *future health* of the client; but it remains to
understand the particulars of this.

To say that the context for prognosis inherently is an "attack" of disease or a
"case" of it is to imply that prognosis is a concept of clinical medicine alone,
exclusive of community medicine. But there undeniably is community prognosis –
and community diagnosis and etiognosis besides. And as for prognosis in clinical
medicine, nothing justifies limiting the concern for the client's future sickness
and/or illness to the context of the client having an existing illness (Sect. 1.4).
For, future sickness and illness are of concern to clinical doctors in respect to
healthy persons as well (Sect. 1.4) – and constitute the entire concern of preventive
medicine in clinical practice.

In the particular context of an existing case of illness, prognosis about this illness
itself concerns its course and complications and not merely of its outcome; the
concern as to outcome is not limited to possible recovery from the illness but
encompasses fatal outcome and survival with sequelae besides; and the illness
itself can be a defect or an injury and not merely a disease, much less merely an
attack-type disease. (Sect. 1.4.)

A clinical doctor's prognosis about a future phenomenon of health that might be experienced by the client is knowing/gnosis about the probability (objective) that it will occur – that the health event at issue will occur in a particular period of prospective time or that the health state at issue will prevail at a particular point in prospective time (conditionally on having survived to possibly experience it). When the probability is taken not to be very high, this perception does not amount to a forecast, a prediction. (Sect. 1.4.)

The probability at issue in prognosis is conditional on the available set of relevant facts – on the prognostic profile of the case at the time of the prognostication – and also a given prospective intervention (preventive or therapeutic) and/or lifestyle, chosen or merely considered, which brings possible adverse effects of the intervention into the set of phenomena to be addressed in prognosis (Sect. 1.4). Thus, even when the context of prognostication is a case of illness, the conditionality of it is not simply "the nature and symptoms of the case," and 'clinical epidemiologists' (e.g., ref. below) are mistaken in their not addressing intervention effects under prognosis.

Reference: Grobee DE, Hoes AW (2009) Clinical epidemiology: principles, methods, and applications for clinical research. Jones and Bartlett Publishers, Boston

Particularly emblematic of the failure to understand prognosis to be a doctor's knowing (about the future health of a client) is the common *attribution of prognosis to an illness* – in saying, for example, that the illness in question 'has a good prognosis.' (Different from a doctor, an illness is not capable of cogitation, prognostic or any other.) Properly understood, good prognosis is correct prognosis – attributing the correct probability to the prospective occurrence in question (Sect. 1.4). Good prognosis is, thus, justified prognosis; different from prediction, prognosis is not subject to post-hoc evaluation on the basis of whether the event/state in question actually did occur.

7.2 The Essence of a Prognostic Study

While the definitional essence of any gnostic study is its being a project to produce evidence for the advancement of the knowledge-base (general) of gnosis (particularistic), a prognostic study naturally is a project to produce evidence for the advancement of the knowledge-base of prognosis. Its objects therefore have bearing on the setting of prognostic probabilities – prognostication – in scientific medicine; they are the parameters in a *prognostic probability function* designed (as to its domain and form) for the study. The evidence from a prognostic study, presented in the report on it, is a set of empirical values (and measures of their imprecision) for the parameters of the PPF at issue – an empirical PPF, that is – together with documentation of how this result was derived.

Paradigmatic of a prognostic study can be taken to be the *randomized trial* of interventions, provided that it is a study not simply of intervention effects but of PPFs for the phenomena of health (sickness and/or illness) in question, and that the choice of intervention is accounted for in the PPFs. Concerning any given one of those phenomena, the effect of an/the index intervention relative to the reference intervention is represented by the difference between the respective probabilities, in terms of which the effect may well be a function of at least some of the prognostic indicators in the PPF, as well as of the prognostic time in it.

What the randomized-trial paradigm, thus extended and specified, teaches about the non-definitional essence – deducible from the definitional essence (above) – of any prognostic study is this: In a prognostic study, the study population is a *cohort* (recruited from some source population and) representing at enrollment (at cohort T_0) the domain of the designed PPFs at issue, that is, the domain in which the prognostication at issue takes place. The cohort is *followed* into prospective cohort-time (which coincides with prognostic time) to implement and document the interventions insofar as they are experimental, otherwise to merely document them (if any); and it is followed to observe and document the prospective occurrence of the health phenomena at issue in the PPFs – with *end-point* of the follow-up for each of the phenomena suitably defined (Sect. 7.5).

From the documented experience is derived not empirical rate-ratio functions as in etiognostic studies but intervention-conditional empirical functions for *rates themselves*, proportion-type rates as measures of probabilities.

7.3   Prognostic Studies' Objects Design

In the objects of prognostic studies there is a *duality* of note. The time-horizon for the prognoses can be very short, as when at issue is a possible acute complication or outcome of an acute illness; or the time horizon is long enough so that conditionality on not succumbing to some extraneous cause of death needs to be an explicit feature of the prognosis (Sect. 2.3). Prognosis about fatal outcome of a case of myocardial infarction is an example of the former, while prognosis about a recurrence of MI generally needs to be explicitly conditional on otherwise surviving the time period at issue in the prognosis.

For inherently very *short-term* prognosis, the occurrence relation addressed in a prognostic study can be formulated directly for the prognostic probability at issue – generally as a linear model for the logit metameter of this probability, a logistic function (Sect. 2.1).

For *long-term* prognosis about an *event*-type phenomenon the need generally is to study directly the incidence density of this event's occurrence as a function of prognostic time (jointly with the prognostic indicators at prognostic T_0 and also the intervention options as of that T_0, if any). Given this ID(t) function, the corresponding prognostic probability function is

$$PPF(t) = 1 - \exp\left[-\int_0^t ID(t)\,dt\right].$$

But if the prognosis is about a *state* of health (sickness or illness), the PPF(t) can be designed for this state's probability directly, just as for either type of phenomenon in the context of very short-term prognosis.

If that integration of ID(t) poses a challenge, one may consider addressing the probability of the event's occurrence before some time $T = t$, expressed by that PPF(t) above, in terms of the probability of *status post* the event at $T = t$. If at issue is death from any cause, these two probabilities are identical by definition, and the prevalence formulation of the function for direct study thus is realistic if there will be no terminations of follow-up before $T = t$ other than those due to death – so that, at $T = t$, each member of the study cohort can be classified as being either alive or in status-post-mortem.

What about prognosis concerning the possible occurrence of a second myocardial infarction in prospective time with T_0 at non-fatal first MI? Even if all members of a cohort formed as of that prognostic T_0 would be followed to $T = t$ or its antecedent death, the cumulative incidence of second MI up to that point in time would not be validly manifest in the prevalence status post this event among the members of the cohort (all of them) at that time. For, now it is not the case that, at any time $T = t$, each member of the study cohort will be classifiable according to presence/absence of the status post second MI – now *conditionally on survival* to that point in prognostic time in the absence of the event at issue. For there will be interim deaths from extraneous causes.

Apart from the exception above, a general basis for the infeasibility of the prevalence approach to cumulative incidence is not only extraneous causes of death but also the 'administrative' terminations of follow-up before $T = t$ (on account of the study's 'common closing date').

While the length of the time horizon in prognosis (very short versus not so) and the nature of the phenomenon at issue – event versus state – thus imply what constitutes the left-hand side of the occurrence relation to be studied (the logit of a probability or the logarithm of the numerical element in incidence density) the real challenge in this function's design only is the formulation of its right-hand side. For the latter, an a-priori given generally is its generic nature as a linear compound of the parameters that constitute the objects of the study (with the designed independent variates, Xs, the coefficients in this compound; Sect. 1.2).

The challenge thus is the specification of the set of independent variates, notably for the first one of the original studies for the prognoses at issue. For any replication of a previous study, needed only is judgment affirming the adequacy of the form of the originally designed PPF.

For the first original study, a set of independent variates invariably needs to be designed for the set of prognostic indicators; one or more other Xs needs to be designed for representation of prognostic time when the time horizon for prognosis is not limited to a very short term; and one or more Xs also needs to be designed for representation of the choice of intervention at prognostic T_0 insofar as such a choice is there in the domain of the prognostication.

An introduction to the design of the objects-defining forms of PPFs, is given in the context of the examples in Sect. 7.5. Suffice it to note here that insofar as intervention effects are involved in the PPFs, the intervention contrast(s) need to involve only candidates for intervention (algorithm) of choice – and that among these in the practice of scientific medicine are not placebo/sham interventions nor the non-algorithmic mélange constituted by 'usual care.'

7.4   Prognostic Studies' Methods Design

The design of the methodology for a prognostic study is, naturally, developed under the constraints of the general essence of these studies (Sect. 7.2) and the study's objects design. The latter defines the entities that need to be operationalized in the methods design: the domain (that of the PPFs) from which the study cohort is to be recruited; the subdomains among which distinctions are to be made within the domain (according to the prognostic indicators, as these are involved in the definitions of the Xs); the interventions (their algorithms) to be effected or initiated as of enrollments into the study cohort and maintained (solo) throughout this cohort's follow-up, insofar as interventions are involved in the PPFs; and the phenomena whose probabilities the PPFs address and which, therefore, are to be documented in the study cohort's follow-up.

As the aim of a study for scientific prognosis is to help advance the scientific knowledge-base of prognostic probability-setting – quantitative – great care is needed in the operational definitions of the entities that are involved. They need to satisfy the requirement of *objectivity* – shared meaning for everyone concerned – and this means the need to be not only clear but also sufficiently *specific*. Otherwise the evidence from the study cannot be correctly and meaningfully interpreted (incl. for replication of the study and for derivative studies on the PPFs at issue).

To satisfy this objectivity/specificity requirement, the enrollments into the study cohort generally need to be *prospective in study time*, with prospective documentation of the domain criteria and the prognostic indicators, for a start.

The methodologic design obviously is more challenging for a study concerned with prognosis beyond the immediate future and directed to an event-type occurrence. The follow-up of the members of the study cohort needs to be consistent with assuring validity. To this end, each surviving member of the cohort needs to be followed to a valid *end point*. The follow-up of any member of the study cohort, when still alive, needs to be terminated as of the time of the occurrence of the event at issue (the cardinal end-point), while otherwise the follow-up is to continue throughout the designed risk period or, possibly, to the study's 'common closing date.' Terminations by loss to follow-up need to be kept to the absolute minimum attainable.

If the PPFs are directed to *intervention*-prognosis, the study generally needs to be experimental and involve *randomization*-based allocation of the study subjects to the compared algorithms of intervention. These allocations need to be followed by close adherence to them, and this means, among other things, the need to select the study subjects with a view to likely attainment of this and the completion of the follow-up. The alternative to this in principle is the corresponding quasi-experiment, in which subject selection replaces the experimental allocation to interventions, and control of confounding (at cohort T_0) replaces its prevention by randomization.

Given all the necessary elementary documentation for each member of the cohort – the prognostic indicators at cohort/prognostic T_0, the subsequent intervention, if any, and the nature of the end point of the follow-up together with the timing of this on the scale of cohort time – the question becomes: How is one to proceed from these data to the study result, to the empirical counterpart of the designed PPF (the design specifying the form of the result, shy of its empirical content)?

Upon the translation of the primary data to the corresponding realizations for the statistical variates involved in a given PPF, at hand is the *case series* of the outcome event with its associated realizations for the dependent variate ($Y = 1$) and for each of the independent variates (Xs), having to do with the prognostic indicators at T_0 and with the event's timing, and commonly with the choice of intervention. The question then is, specifically, about transition from these data to

$$\log\left(\mathrm{ID}'\right) = \hat{B}_0 + \sum_i \hat{B}_i X_i,$$

where ID' is the numerical element in the ID at issue (number of events per unit amount of population-time) and the right-hand side involves the empirical values for the parameters constituting the objects of the study.

The thus-documented case/numerator series needs to be supplemented by an analogously documented *denominator series* (for which $Y = 0$), selected as a *representative sample* of the population-time of follow-up relevant to the PPF at issue, and the size of this population-time needs to be determined. Then, the logistic model

$$\log[\Pr(Y = 1)/\Pr(Y = 0)] = B_0 + \sum_i B_i X_i,$$

needs to be fitted to the data on the pair of series (of persons-moments) and translated into

$$\widehat{ID}(t) = (B/B*)\exp\left(\hat{B}_0 + \sum_i \hat{B}_i X_i\right),$$

where t is represented by its corresponding statistical variate(s); B and B* are the respective sizes of the denominator series and the denominator population-time itself; and the parameters' empirical values are the ones obtained from the fitting of the logistic model. And finally,

$$\widehat{PPF} = 1 - \exp\left[-\int_0^t \widehat{ID}(t)\,dt\right].$$

Prognostic studies are, most commonly by far, experiments in which the intended effects of two or more intervention options are compared, the intention with the interventions being to reduce the probability (risk) of an adverse health event over a considerable span of prospective time following the interventions' initiations. Intervention-conditional PPFs generally are not addressed at all, and the effects of the interventions thus are not assessed in terms differences in the probabilities implied by those PPFs – the way the difference is a function of prognostic time conditionally on a given prognostic profile at prognostic T_0. Instead, the interventions now usually are compared in terms of a single-valued '*hazard ratio*' (indidence-density ratio) derived under the 'proportional hazards' model of D. R. Cox.

Contemplation of intervention-prognosis in terms of knowledge about this HR does not conform to the rationality requirements of scientific medicine.

7.5 Some Recent Studies Critically Examined

The second one of the two example studies on original diagnostic research in Sect. 5.10 addressed a novel, non-invasive test for diagnosis about the causal role of a coronary stenosis in the angina pectoris of chronic coronary heart disease. The interest in this test was inspired by the results of a study on the intervention-prognostic implications of the corresponding invasive test, addressed in a randomized trial that had been published in two stages:

References:

1. Tonino PAL, De Bruyne B, Pijls NHJ et al (2009) Fractional flow reserve versus angiography for guiding percutaneous coronary intervention. N Engl J Med 360:213–224

2. Pijls NHJ, Fearon WF, Tonino PAL et al (2010) Franctional flow reserve versus angiography for guiding percutaneous coronary intervention in patients with multivessel coronary artery disease. 2-year follow-up of the FAME (Fractional Flow Reserve Versus Angiography for Multivessel Evaluation) study. J Amer Coll Cardiol 56:177–184

This trial is here addressed as the example *study 1*, focusing on the 1-year follow-up report (ref. 1 above).

According to that first report on the study, "The objective of this randomized study was to compare the treatment based on the measurement of [the proportional pressure gradients across the stenoses] with the current practice of treatment guided solely by angiography in patients with multivessel coronary artery disease for whom PCI ["percutaneous coronary intervention," meaning coronary angioplasty as a matter of inserting a drug-eluting stent] is the appropriate treatment." The comparison had a 1-year time horizon for the relative effects. The second report said that "The purpose of this study was to investigate the 2-year outcome of [PCI] guided by [that measurement] in patients with multivessel coronary artery disease (CAD)."

The rationale for addressing this contrast was implied by the statement that, "For patients with multivessel coronary artery disease, [such selective use of stenting], while still achieving complete relief of myocardial ischemia [angina, that is], could improve the clinical outcome and decrease health care costs" (ref. 1 above). Implied by this is that the sole purpose of stenting is to obtain relief from angina, and that other possible effects likely are adverse on balance. An added implication is that, despite those (unspecified) concerns about adverse side effects, angina-relieving stenting is preferable to its alternatives. No explication of either one of these implications was given in either one of the two reports.

The rationale for the study would have been explicit in the reports had the authors shared with their readers the information they gave to their patients when soliciting them to sign a "written informed [*sic*] consent" to becoming subjects in the study. But, as is routine in the prevailing culture of even these experimental intervention-prognostic studies on patients, this information was not given in either one of the two reports on the study.

Had that information been given, interesting to a critical reader would have been to learn, first, what was said to the patients about the adverse effects of the pressure-gradient measurement and of the stenting itself; and what was said about different but reasonable ways for the patients to think, in the context of the given information and their personal valuations, about the relative merits of the stenting and its alternatives (medical and bypass-surgical).

And particularly interesting, both scientifically and with a view to ethics, would have been to learn whether the patients were told that the investigators know who are the "patients with multivessel coronary artery disease for whom PCI is the appropriate treatment" (cf. above).

Free from that hubris, designed for the study – for its objects and thereby for the study proper – would have been a domain in which stenting is an option to consider and purely medical treatment is the (only) alternative to this. To this end designed should have been, first and foremost, the domain-definitional particulars of the chronic CHD (coronary heart disease) as to the nature, severity, and promptings of the angina and also the domain-definitional patterns of the angiographic findings on the coronaries. Other definitional features of the domain might well have been designed to be history of no acute CHD and no previous coronary angioplasty, among other features.

In reference to the designed domain of the study, the need was to design the form of the prognostic probability function for the entity the occurrence and/or severity of which the interventions are intended to favorably affect; and in addition, it was to be designed for at least the entity that is of principal concern regarding the interventions' possible adverse effects. This means the need to design (the form of) the PPF for exertion-induced angina and for acute CHD at least.

In these designs, nothing is served by designating one of the entities as being of 'primary' concern and the other(s) as being (only) of 'secondary' concern. In particular, the weight of the evidence the study will produce in respect to a given one of the entities will in no way be dependent on these designations, routine though they are in the prevailing culture of intervention-prognostic studies, experimental ones in particular, by demand of journal editors.

This prevailing culture cannot be simply ignored as irrelevant, for it seriously distorts prognostic studies' objects designs and thereby the kinds of result that are reported from them. When the culture-bound idea is that a single entity is to be designated as being of 'primary' prognostic interest in any intervention-prognostic study, this commonly has the consequence that various entities that in practice are addressed separately are in research aggregated into a single, composite one. Thus, said in the first one of those two reports was this:

> The primary end point was the rate of major cardiac events at 1 year. Major cardiac events were defined as a composite of death, myocardial infarction, and repeat revascularization.

Doctors and patients are prone to have great difficulties with the knowledge derived from research on such composite 'end points' (which should be construed as events per se, not rates of the occurrence of these; cf. Sect. 7.4 above). Thus, when a patient needs to choose between angioplasty (with medications) and purely medicational intervention and asks to be informed about his/her doctor's science-based 1-year prognosis about this in respect to fatal myocardial infarction, prognosis about this specific to his/her particular case of chronic CHD and to the two possible choices of intervention, the patient would be astounded by the doctor's correct answer.

The doctor would have to say, for a start, that medical science for patients with chronic CHD does not address prospective happenings as specific as death from myocardial infarction; that it does not specifically address even death per se, regardless of its proximal cause; that it addresses a composite that is not even restricted to actual entities of health; that for cases of chronic CHD it involves one aspect of possible extraneous healthcare, namely repeat revascularization, without any reference to the indications for or nature of it.

And the doctor would have to add that not only does the prevailing, presumedly practice-serving science not address, separately, the prospective occurrence and/or severity of such specific entities of health as actually are of prognostic concern to both doctors and their patients; the composite entity it contrives to address it addresses without specificity to individual patients' particular prognostic profiles and without specificity to the intervention (comprehensive stenting, limited stenting, or no stenting) that also would be predicates in whatever reasonable prognosis.

The doctor would have to say, in short, that in the framework of the culture that now surrounds intervention prognostic research, no science-based answer to the very reasonable prognostic question (above) can be given, not now nor in the future so long as this culture will continue to prevail.

Upon the requisite reformation of the culture of intervention-prognostic medical research, the counterpart of this study would be quite different.

For the domain (sketched above) of patients with chronic CHD who are considering intervention by one or the other of the two types of coronary angioplasty (above) or neither, this in the context of being about to undergo coronary angiography, one type of original (or derivative) study would address a predesigned prognostic probability function for exertion-induced angina, another for acute CHD; and yet other PPFs might address the patients' future health in yet other respects.

The designing of these PPFs would be done before any of the studies on them; and it would be done with great care. For, these objects designs would be successful only insofar as there would be agreement about their appropriateness in the relevant scientific community, so that various teams of investigators would be willing to study them, unmodified. The original designs of those PPFs (their forms) would be published and publicly discussed, with consensus about them developed through this process. In all of this, the PPFs would be addressed in purely conceptual terms first; and then, subordinate to this, operationally. (The latter might call for modification of the former.)

In the methods designs of these studies, a novelty of note beyond the particular topic of practice that is at issue here would be the absence of 'statistical methods' and their subtopic of 'sample size determination.' For, the methods designs would cover the fitting of the predesigned PPFs to the data, and any given team of investigators would make a contribution – ultimately to derivative studies – of whatever degree of informativeness/precision it is willing and able to provide. There thus would be no place for ideas such as, "The estimated minimum sample size of 426 patients in each group was based on a two-sided chi-square test with an alpha level of 0.05 and a statistical power of 0.80, assuming 1-year rates of major cardiac events of 14% in the [inclusive stenting] group and 8% in the [select stenting] groups" (ref. 1 above).

The methodology of each of these studies would be designed and executed in objective terms. There thus would be no place for post hoc descriptions such as, "the investigator indicated which lesions … were thought [*sic*] to require PCI on the basis of angiographic appearance and clinical data" (ref. 1). The objective counterpart of this would be defined by the operationalized versions of the designed objects of study (cf. above).

The choice among the three types of possible intervention would be randomized (as it was between the two of these in the study reported in refs. 1 and 2 above), but it also would be 'blinded' to both the patients and the investigators involved in the follow-up. For this would not only be feasible; it would help objective implementation of the protocols for documentations and interventions during the follow-up (identical in algorithms among the three subcohorts defined by baseline intervention); and it would be particularly relevant for comparability, across the treatment groups, of the (subjective) reports on the level of angina at various times in the course of the follow-up.

Objectivity – universality/singularity of meaning for all qualified readers – would characterize everything in the report of each of the studies. Each report thus would be free of statements such as, "The primary end point was the rate of death, nonfatal myocardial infarction, and repeat revascularization at 1 year" (ref. 1 above). The meaning of this must be something other than the literal one; for meant could not have been these events occurring "at [*sic*] 1 year" (of follow-up). As for these events occurring by – before – 1 year of follow-up, is the intended meaning the literal one, namely that "the primary end point" was the rate of all three events occurring before 1 year of follow-up? Or was it, instead, that of at least one of them occurring, in which case the needed conjunction would have been 'and/or' and not 'and'? Whatever may have been the event in question, is one really to believe that it wasn't this event per se but its rate – undefined – of occurrence that the authors took to be the general (objective) meaning of 'end point' in prognostic studies? And whatever may have been the 'end point,' in what meaning was it "primary"?

As a second example of recently reported original prognostic studies – out of a total of eight – I take up a purely descriptive-prognostic one, as distinct from the intervention-prognostic study addressed above. Like that study 1, this *study 2* also addressed risk of cardiovascular disease:

Reference 3: Yeboah J, McClelland RL, Polonsky TS et al (2012) Comparison of novel risk markers for improvement of cardiovascular risk assessment in intermediate-risk individuals. JAMA 308:788–795

The report's introductory section converged to this phrasing of the study's generic object:

> In this report, we assess the improvements in prediction accuracy and reclassification to high- and low-risk categories using [five laboratory-based prognostic indicators] and family history of CHD [coronary heart disease] in asymptomatic adults classified as intermediate risk (FRS [Framingham risk score] standards) who participated in the Multi-Ethnic Study of Atherosclerosis (MESA).

In the abstract, the "Objective" of the study was said to have been this:

> We compared improvement of prediction of incident CHD/cardiovascular disease (CVD) of these 6 risk markers within intermediate-risk participants (FRS > 5 % − < 20 %) in the Multi-Ethnic Study of Atherosclerosis (MESA).

Cardiovascular risk was the overaching concern in this study according to the report's title; and as for what this risk specifically was about, the "Main outcome measures" were said to be "Incident CHD defined as myocardial infarction, angina followed by revascularization, resuscitated cardiac arrest, or CHD death. Incident CVD additionally included stroke or CVD death." Ignoring the questionable details in this, let us take it that this study was about the risk – probability of future occurrence – of atherothrombotic cardiac and cerebral disease, separately and combined.

Risk assessment, to which the title refers, in this context is quantification of the risk, estimation of the level of that risk for one or more specified spans of prospective time; in other words, it is development of prognosis about the health event at issue conditionally on the person's prognostic profile – and in general also, though not in the example, conditionally on select options for prospective intervention.

Research for this risk assessment concerning a given one of those two types of event addresses different pre-designed prognostic probability functions, which for a given domain vary mainly according to the prognostic indicators that are accounted for in the prognostic profiles, the other source of variation being the way in which a given indicator is scaled and then entered into the PPF.

In the study at issue here, one of the "novel" indicators of the risks of prognostic interest was *family history* of CHD (cf. above). "Family history was obtained by asking participants whether any member of their immediate family (parents, siblings, and children) experienced fatal or nonfatal myocardial infarction," and "Family history was entered into models as a categorical variable (yes/no)."

A more reasonable approach to this prognostic indicator would have involved, for a start, a bi-phasic questioning. The first question should have been, How many first-degree relatives – parents, siblings, and children – of yours have reached age 60 (say) or died of myocardial infarction before this age? And the second, subordinate question would have been, Of these relatives of yours, how many have had a myocardial infarction (fatal or nonfatal)? And then, given the respective answers N and n, the family history's simple representation in the PPF could have been in

terms of two variates: an indicator for N > O together the product of this with
(n + 1)/(N + 2), say.

Remarkable about the six "novel" indicators of the risks also was that they did
not include ethnicity – in a study that had "multi-ethnic" in its very name.

And whatever was the set of prognostic indicators, particularly notable about
this study was that it did not produce an empirical counterpart of a pre-designed
theoretical PPF. Instead, an example of its results was that "coronary artery calcium
[was] independently associated with incident CHD in multivariable analyses
(HR 2.60 [95% CI, 1.94–3.50])," where HR stands for "hazard ratio," meaning
incidence-density ratio.

This study, alas, did not appreciably advance the evidence for the knowledge-
base of the prognoses in question.

As the third example of original prognostic studies I take up a non-randomized
intervention-prognostic study that was a topic in Sect. 3.3 already. For this *study 3*
the report was this:

Reference 4: Sjöström L, Peltonen M, Jacobson P et al (2012) Bariatric surgery
and long-term cardiovascular events. JAMA 307:56–65

The essence of this study was manifest in its core result (as presented under
"Conclusion"): "Compared with usual care, bariatric surgery was associated with
reduced number of cardiovascular deaths and lower incidence of cardiovascular
events in obese adults."

"The rationale for [this] study was to fill the knowledge gap regarding the
association between type of treatment (bariatric surgery vs. usual care) and hard
end points (primarily mortality). For ethical reasons related to the high postoperative mortality in the 1980s, a randomized design was not approved and a matched
study was therefore undertaken [refs.]."

Implied by this are two untenable ideas: that when random allocation of suitably
informed volunteers to an intervention is deemed to be unethical, it still is seen to be
ethical to otherwise provide this intervention and thereby to make possible this
non-randomized study on it; and that the non-experimental alternative to
experiments' randomization is matching (which actually corresponds to 'blocking'
in randomized experiments and is not necessary for non-experimental assurance of
un-confounded results).

"Recruitment campaigns were undertaken in mass media and at 25 public surgical
departments [*sic*] and . . . Among the eligible patients [among the respondents, that is],
2010 individuals electing surgery constituted the surgery group and a contemporane-
ously matched control group of 2037 participants was created by an automatic
matching program using 18 matching variables (Table 1)." The information on the

basis of which the study participants chose between surgery and "usual care" the report did not reveal. Said was, simply, that "All regional ethical [*sic*] review boards in Sweden approved the study protocol, and all patients gave informed consent to participate." From this it is not clear whether "the study protocol" that was approved included the information presented in the solicitation of some of the respondents to undergo the surgery (at a time when randomized allocation of informed volunteers to the surgical intervention was deemed unethical; cf above).

Relevant for obese persons' suitably informed acceptance of bariatric surgery – specifically one of the three subtypes of it – in preference to non-surgical care is, for one, information about surgical fatality risk, specific to the type of surgery. Remarkably, nothing about this was included in the reported results.

As for what else may be relevant, the "Objective" of this study was "To study the association between bariatric surgery, weight loss, and cardiovascular events," specifically myocardial infarction and stroke. The results had to do with "hazard ratios" and "interactions." Presented in three tables were 117 P-values and 239 imprecision intervals, supplemented by others in the three figures.

Said under the "Comment" was that "The main limitation of [this] study is that the intervention was not randomized, and this was due to the high postoperative mortality in the 1980s [ref.]." I say, by contrast, that a major limitation of this study was that it did not address any prognostic probability function.

The next example here is *study 4*:

Reference 5:   De Berardis G, Lucisano G, D'Ettorre A et al (2012) Association of aspirin use with major bleeding in patients with and without diabetes. JAMA 307:2286–2294

In this study, "Individuals with new prescription for low-dose aspirin ($\leq$300 mg) were identified ... and were propensity-matched on a 1-to-1 basis with individuals who did not take aspirin [in the same population in the same period of calendar time]. .... During a median follow-up of 5.7 years, the overall incidence rate of hemorrhagic events was 5.58 (95% CI, 5.39–5.77) per 1000 person-years for aspirin users and 3.60 (95% CI, 3.48–3.72) per 1000 person-years for those without aspirin use (incidence rate ratio [IRR], 1.55; 95% CI, 1.48–1.63). .... Irrespective of aspirin use, diabetes was independently associated with increased risk of major bleeding episodes (IRR 1.36; 95% CI, 1.28–1.44)."

So, this was another *quasi-experimental intervention-prognostic study*, one that focused on an unintended, adverse effect of the intervention relative to no intervention.

The description of the "Study design" was, in part, quite confusing. Thus, "Current low-dose aspirin users were defined as those who had the last prescription for aspirin at least [*sic*] 75 days before hospitalization for major bleeding events or the end of follow-up; this allowed a tolerance of 15 days between prescriptions, each covering a period of 60 days of treatment. .... Former aspirin users, who were those who had prescriptions for aspirin at the beginning of follow-up but had their last prescription of aspirin more than 75 days before an event, were excluded [*sic*] from the analyses."

But let us take it, for the purposes here, that the authors identified – in the time-course of a very large dynamic source population – people who started regular use of 'low-dose' aspirin, and that the investigators followed them to the time of the discontinuation of this regimen (with no major bleed), occurrence of a major bleed, death from an extraneous cause, loss to follow-up, or the study's 'common closing date,' whichever came first. And let us take it that they also identified and analogously followed persons who also initially did not use aspirin in the period of the source population's follow-up, with the discontinuation of no regular use the counterpart of discontinuation of regular use, except that the non-users' follow-up as to its inceptions and endings was truncated to the range of age that characterized the user cohort's follow-up. Let us take it that the study involved these two cohorts' follow-up to those end-points of it.

Before dealing with the risk of the bleeding event, the study needs to address this event's *incidence density* in that population-time of follow-up. This ID indeed was the authors' concern just the same (cf. above), though they did not proceed from this to dealing with any risk function.

For this first stage of the study, the objects design should have resulted in something like this (as the definition of the parameters of interest):

$$\log\left(\text{ID}'\right) = B_0 + \sum_1^7 B_i X_i,$$

where ID' is the numerical element in the index ID, and the independent variates in this are:

X_1:   indicator of regular 'low-dose' aspirin use (1 if use, 0 otherwise);
X_2:   dose of regular 'low-dose' aspirin use (number of mg/day; 0 if $X_1 = 0$;
X_3:   age at cohort T_0 (at initiation of regular regimen of aspirin use/non-use; number of years);
X_4:   duration of regular aspirin use/non-use (number of years);
X_5: $X_3 X_4$;
X_6:   indicator of diabetes (1 if diabetes, 0 otherwise); and
X_7: $X_1 X_6$.

To document this function for the population-time of follow-up, the cases of the event occurring in that population-time would need to be identified; and this case series would need to be supplemented by a *representative sample* of the (infinite number of) person-moments constituting the population-time of follow-up. The two series, with $Y = 1$ and $Y = 0$, respectively, would be documented in respect to those independent variates. Then, the logistic model,

$$\log[\Pr(Y = 1)/\Pr(Y = 0)] = B_0 + \sum_1^7 B_i X_i,$$

would be fitted to the data. This would allow documentation of the ID in the study base as:

$$\text{ID} = (B/B^*)\exp\left(\hat{B}_0 + \sum_1^7 \hat{B}_i X_i,\right),$$

where B is the size of the Y = 0 series and B* is the amount of population-time of follow-up.

Given that risk over a span of prospective time from T = 0 to T = t, in number of years, is

$$R_{0,t} = 1 - \exp\left(-\int_0^t \text{ID}_t \, dt\right),$$

those ID integrals would need to be derived from the ID function above, separately for users and non-users of 'low-dose' aspirin. The result for the users would be

$$\hat{R}_1 = (B/B^*)\exp\left[\hat{B}_0 + \hat{B}_1 + \hat{B}_2 X_2 + \hat{B}_3 X_3 \right.$$
$$\left. +(\hat{B}_6 + \hat{B}_7)X_6\right]\int_0^t \exp\left[(\hat{B}_4 + \hat{B}_5 X_3)X_4\right] dX_4,$$

in which

$$\int_0^t \exp\left[(\hat{B}_4 + \hat{B}_5 X_3)X_4\right] dX_4 = \{\exp[(\hat{B}_4 + (\hat{B}_5 X_3)t] - 1\}/(\hat{B}_4 + \hat{B}_5 X_3).$$

For non-users, correspondingly, the result for the t-year risk is

$$\hat{R}_0 = (B/B^*)\exp\left(\hat{B}_0 + \hat{B}_6 X_6\right)\int_0^t \exp\left(\hat{B}_3 X_3\right) dX_3$$
$$= (B/B^*)\exp\left(\hat{B}_0 + \hat{B}_6 X_6\right)\left[\exp\left(\hat{B}_3 t\right) - 1\right]/\hat{B}_3.$$

The prevailing culture in non-experimental research on the effects of treatments and prophylactics is well illustrated by two quite similar studies, *study 5* (ref. 6 below) and *study 6* (ref. 7):

References:

6. Svanström H, Pasternak B, Hviid A (2013) Association of treatment with losartan vs candesartan and mortality among patients with heart failure. JAMA 307:1506–1512
7. Zhang J, Xie I, Delzell E et al (2012) Association between vaccination for herpes zoster and risk of herpes zoster infection among older patients with selected immune-mediated diseases. JAMA 308:43–49

In these studies, the investigators' background in etiogenetic/etiognostic research is evident. For a start, no report from an experimental study on the effects at issue is described as addressing "associations" between phenomena of health and their antecedent treatment or prophylaxis (cf. refs. 6, 7 above). And both of these quasi-experimental studies were characterized as "cohort studies" in line with the prevailing (though untenable) taxonomy of etiogenetic/etiognostic studies, while the experimental counterparts of these would never be characterized as cohort studies.

A causal study of this, prognostic genre, distinct from an etiognostic one, should address risks; and said in the report of each of these two studies indeed was that risk was addressed, with "risk of all-cause mortality" (ref. 6 above) presumably meaning risk of death from any cause. But: actually addressed in both studies was effect on incidence-density rather than cumulative incidence (which implies risk).

This ID, even, was addressed in extremely simplistic terms. For example, for patients hospitalized for congestive heart failure, but otherwise specified only by minimum age, "Among 4397 users of losartan [dosage unspecified], 1212 deaths [of any cause] occurred during 11347 person-years of follow-up [of unspecified durations] (unadjusted incidence rate [IR]/100 person-years, 10.7; 95% CI, 10.1–11.3) compared with . . ." (ref. 6). And from the other study (ref. 7), analogously reported were "Herpes zoster incidence rate within 42 days after vaccination (safety concern) and beyond 42 days" as the number of cases per 1,000 person-years, and for the unvaccinated as well. From both studies, incidence-density ratio ("hazard ratio") with adjustment for some potential confounders was derived using Cox regression.

Both of these studies were characterized as ones of hypothesis testing. Nevertheless, as is usual and proper, quantitative results were reported. But, as also is usual yet very improper, the form – including extreme non-specificity – of those results was greatly at variance with the form of the requisite knowledge-base of intervention-prognosis.

This failure to design a meaningful form for the objects (and thereby the results) of an intervention-prognostic hypothesis-testing study comes to particularly poignant focus in the example study addressed next.

The seventh example here is a very major randomized trial, having to do with intervention-prognosis about fatal outcome of lung cancer. This *study 7* contrasts screening-afforded early treatment with the late treatment that occurs in the absence of screening:

References:

8. Statement concerning the National Lung Screening Trial. Oct 28, 2010. http://www.cancer.gov/images/DSMB-NLST.pdf

9. The National Lung Screening Trial Team (2011) Reduced lung-cancer mortality with low-dose computed tomography screening. N Engl J Med 365:395–409

"The National Lung Screening Trial (NLST) was conducted to determine whether screening with low-dose CT [computed tomography] could reduce mortality from lung cancer. . . . [W]e enrolled 53,454 persons at high risk for lung cancer. . . . Participants were randomly assigned to undergo three annual screenings with either low-dose CT (26,722 participants) or single-view posteroanterior chest radiography (26,732). . . . There were 247 deaths from lung cancer per 100,000 person-years in the low-dose CT group and 309 deaths per 100,000 person-years in the radiography group, representing a relative reduction in mortality from lung cancer with low-dose CT screening of 20.0% (95% CI, 6.8 to 26.7; P = 0.004). . . . Conclusion: Screening with the use of low-dose CT reduces mortality from lung cancer." (Ref. 9 above.)

That result of the NLST, for its hypothesis testing, has been taken to have great quantitative significance, including by eminent medical organizations. Thus, the American College of Chest Physicians and the American Society of Clinical Oncologists in their guidelines for doctors in their advising of people on CT screening for lung cancer (refs. 8, 9 below) take that result to mean that CT screening – really its associated early treatment – reduced, and hence generally reduces, the *case-fatality rate* of lung cancer by 20.0 %. For they say that "4 out of 5 people who are going to die of lung cancer will die of it even if they are screened. Screening prevents one in five deaths from lung cancer."

References:

10. Bach PB, Mirkin JN, Oliver TK et al (2012) Benefits and harms of CT screening for lung cancer: a systematic review [published online ahead of print May 20, 2012]. JAMA. doi: 10.1001/jama.2012.5521
11. Bach PB, Mirkin JN, Oliver TK et al (2012) The role of CT screening for lung cancer in clinical practice. The evidence based practice guideline of the American College of Chest Physicians and the American Society of Clinical Oncology. In eAppendix 4. Benefits and harms of CT screening for lung cancer: a systematic review [published online ahead of print May 20, 2012]. JAMA. doi: 10.1001/jama.2012.5521

While reduction of the probability of dying from lung cancer (per its reduced case-fatality rate) indeed is the aim in seeking screening for this cancer, the NLST was not designed to quantify this reduction (cf. above). The *design of the NLST* was in line with the design of randomized trials on screening for a cancer in general. The general ideas in the trial design are these: The screening is a single test (rather than a diagnostic algorithm) constituting an intervention (with the aim of preventing death from the cancer). The intended consequence of screening for a cancer, like that of any medical intervention, is best assessed by means of an intervention trial. In the trial, the subjects are randomly assigned either to the testing and, in case of its negative result, periodic repetitions of this up to a designated (small) number of times, or to refraining from the testing. Each subject is followed up to the first one of: death from the cancer, death from an extraneous cause, and the trial's 'common closing date.' For each of the two arms of the trial, the *core result* is the number of deaths from the cancer in unit amount of population-time of follow-up (cf. core results of the NLST above).

This discrepancy between what is needed for practice and what actually is addressed in trials such as the NLST, raises a very important question: How should the trial design be modified so as to make it into a genuine *intervention-prognostic study*, addressing diagnosis-conditional risk of dying from the cancer and contrasting, in this, early, screening-associated treatment with treatment that is late on account of diagnosis in the absence of any screening?

In answering this question I leave aside the issues of diagnostic algorithms and treatment protocols, which should be carefully designed but commonly are not. I focus on the trial design in respect to: a needed restriction in the randomization; the proper individual durations of follow-up; the needed further documentation of the cases of death in the trial's screening arm; and the appropriate way to synthesize the data – for a result on the proportional reduction in the case-fatality rate resulting from screening-associated early treatments in place of late treatments in the absence of any screening.

The randomization in the context of equal allocation (as in the NLST) could be done with restriction to '*blocks' of two* subjects, with one member screened (with a designed algorithm for the pursuit of the diagnosis) and the other member not screened; and the *maximum, default duration of follow-up* – the designed duration in the absence of death before it – for both members of the pair should be long enough but not too long. This default duration of follow-up (for death from the cancer) should be designed to be long enough to cover the maximum time from screening-afforded cure (by early treatment) to the death that thus gets to be prevented but with the understanding that follow-up longer than this, while also consistent with validity, takes away from the result's precision and thus is to be avoided. *Blocks of three* would be preferable, with one screened and the other two not screened, as the efficiency-optimal allocation ratio is inversely proportional to the square root of the unit cost ratio.

In the deaths from the cancer in the trial's screening arm, *a distinction is to be made* according to whether the detection of the cancer was accomplished under the screening – by the screening or on account of symptoms-prompted diagnostics between scheduled screenings – or only after its discontinuation. The deaths from cases diagnosed under the screening represent failures to achieve cure by early treatment, while those from cases diagnosed after the screening's discontinuation have no bearing on the merits of the screening. This means the need to secure these data and then to translate the result for proportional reduction in all of these deaths to the counterpart of this specific to the cases that were diagnosed under the screening.

In these terms, while there was a total of C cases of death from the cancer in the trial's screening arm, C_1 of these were identified to be from cancers diagnosed under the screening and C_0 from subsequently diagnosed cases; $C = C_1 + C_0$. And based on the trial's control arm, the 'expected' number corresponding to C was C', so that the result for the proportional reduction in these deaths was $P = (C' - C)/C' = 1 - C/C'$; $C = C'(1-P)$. The question is, what therefore is the corresponding $P_1 = (C'_1 - C_1)/C'_1$?

This desired result can be deduced from these elementary ones as follows:

$$C_1 + C_0 = C;$$

$$C'_1(1P_1) + C_0 = C'(1P);$$

and since $C'_0 = C_0$ and $C'_1 = C'_0 = C'$, we have

$$C'_1 P_1 = C'P;$$

$$[C_1/(1 - P_1)]P_1 = [C/(1 - P)]P;$$

$$P_1/(1 - P_1) = [P/(1 - P)]/(C_1/C).$$

Thus, for example, if $P = 0.050$ and $C_1/C = 0.053$, the result for $P_1/(1 - P_1)$ is 1.0 and that for P therefore is $0.5 = 50\%$ – so that the result for case-fatality reduction is greater by an order of magnitude relative to the result for the reduction in mortality rate.

If the designed default duration of follow-up (above) is shorter than what is needed for valid assessment of the reduction in case-fatality rate, the result for this reduction has a downward bias, while an unduly long follow-up reduces the proportional reduction in the mortality rate without thereby biasing the result for the reduction in case-fatality rate; only the precision of the latter would be compromised by unduly long follow-up, due to increased relative imprecision in the result for the reduced proportional reduction in the mortality rate.

Not only can the intended consequence of screening for a cancer in principle be assessed by means of a randomized trial with these modification of its design; quantifiable also becomes the principal adverse consequence that the screening could have. This adverse consequence is the propensity of the diagnoses under the screening to be, to some extent, overdiagnoses in the meaning of leading to overtreatments, to treatment as life-threatening lesions ones that are not malignant in this sense.

The overdiagnoses can be quantified this way: When the case-fatality result is that C_1' deaths from cases diagnosed under the screening were reduced to C_1 cases, to yield $P_1 = 1 - C_1/C_1'$, the implication is that all of the C' deaths, including the $C' - C_1$ ones, were from genuinely life-threatening cases. Thus, if a total of C_1'' cases were diagnosed under the screening, the result for the proportion of overdiagnoses among these diagnoses is $(C_1'' - C_1')/C_1'' = 1 - C_1'/C_1''$.

These needed modifications of NLST-type trials on screening for a cancer, in order that they address the reduction in case-fatality rate, raise the question about *possible alternative types of study* to the same end. This interest in possible alternative types of study arises because the NLST, as it was, had a cost as high as U.S. \$250 million, and the needed modifications of it would have added quite considerably to its cost. The added costs would have arisen, in part, from substantial extensions of the durations of follow-up (those of the shortest ones in particular to the needed duration, rather than to the trial's 'common closing date'); but the principal basis for added costs would have been these extensions' consequent need for an increased size of the trial cohort (cf. above), as concern for statistical 'power' for hypothesis-testing is replaced by that for the precision of the result on the reduction in mortality in the particular, relevant meaning of proportional reduction in case-fatality rate.

The *simplest* and by far *the least expensive* alternative to consider is a study focusing on documentation of the case-fatality rate in a cohort based on instances of the cancer's treatment immediately following its diagnosis under the screening (with both the pursuit of the diagnosis and the treatment specified by the study protocol and admissibility to the cohort restricted by this). The cumulative incidence of death from the cancer (post cohort enrollment) reaches an asymptote once the follow-up is long enough (cf. above), and this is the case-fatality rate for the cohort. Reduction in the case-fatality rate arising from the early, screening-associated treatments in place of the late treatments in the absence of any screening is quantified by contrasting this directly documented case-fatality rate with the known counterpart of this in the absence of any screening.

What about the validity of this study? In addressing this question it needs to be appreciated that the reduction in case-fatality rate from screening-associated early treatment must be quite substantial for the screening to be justifiable; that small biases really are not consequential, in proportion to the difference of practical interest. And it is good to bear in mind that the mortality reduction results of trials on screening for a cancer are now regarded as valid measures of case-fatality reduction, despite the huge (downward) biases generally involved in this (cf. above).

If the contrast were to address rates of death from the cancer within, say, 5 years from the time of diagnosis, the study would be marred by substantial 'lead time' bias. For, the early-diagnosed cases in the early-treatment cohort typically take a longer time after diagnosis to manifest their fatal outcome than do cases diagnosed late, in the absence of screening, even if the early treatment offers no curative/survival advantage. But the design at issue here is not about anything like 5-year rates of death/survival, and the consequence of the lead time from screening is merely the need for longer follow-up to attain the asymptote of the cumulative incidence of death from the cancer, representing the case-fatality rate in the study cohort.

A special case of the lead-time issue arises from the propensity of the baseline round of screening to detect cases that are relatively slowly-growing, cases that therefore tend to have their fatal outcomes only after a relatively long time after the diagnosis. The upshot of this is, first, that the baseline-diagnosed cases have an important role in defining the necessary duration of the cohort's follow-up; and second, that it would be good to distinguish between the baseline and repeat rounds of the screening and to form, on this basis, two subcohorts for the assessment of their respective, perhaps appreciably different, case-fatality rates.

The validity problem that does require a special solution – a partial one at least – for this (high-economy) design of the study is *overdiagnosis*, while this is a non-issue for the (enormously expensive) suitably modified screening versus no screening trial (cf. above). This need is inherent in the mission to quantify the case-fatality rate specific to the genuinely life-threatening – truly malignant – ones among all of the lesions diagnosed as representing cancer in the study cohort.

The concern about screening-associated overdiagnosis is practically confined to an identifiable subtype of the lesion diagnosed as cancer in this context. In screening for breast cancer it is the 'ductal carcinoma in situ'; and in CT screening for lung cancer it is 'adenocarcinoma' manifest in the images as a subsolid lesion, common in the baseline round of the screening but uncommon in annual repeat rounds of it.

Given the suspicion of possible overdiagnosis (pathological) in the case of a particular type of lesion, routine resection of the lesion is a questionable practice. A reasonable alternative to consider, in discussion with the person in question, is 'watchful waiting' to see whether the lesion actually is destined to exhibit malignant-type progression if left untreated and to treat it only if and when this propensity gets to be established. Upon such a discussion, some people choose undelayed treatment while others opt for the watchful waiting approach; and those in the latter subcohort provide a result on the proportion of overdiagnosed cases in that proportion of the diagnosed cases in which overdiagnosis is an issue.

All in all, thus, there is a very strong case for a simple, inexpensive, non-experimental alternative to the enormously expensive screening versus no screening trial to assess the reduction in case-fatality rate of a cancer when treated early upon diagnosis under a particular regimen of screening for it. A serous re-examination of this very important topic is eminently called for on the part of agencies such as the ones that bore the cost of the NLST.

The gargantuan intervention-prognostic study (NLST) addressed as example study 7 above was sponsored by the National Cancer Institute of the U.S. together with the American Cancer Society; and very instructive of how the screening experts in the ACS think about the result of that study is their "study" on the implications of the NLST. It was published in the journal of the ACS (ref. below), and I examine it as the example *study 8* – the final one of these here.

Reference 12: Ma J, Ward EM, Smith R, Jemal A (2013) Annual number of lung cancer deaths potentially avertable by screening in the United States. Cancer 119:1381–1385

These authors took it as a given that such screening as was applied in the CT arm of the NLST reduces mortality from lung cancer by 20 % in all of the types of person that were admitted to that trial, while to them it remained "unclear whether screening also is effective among current or former smokers with a smoking history of <30 pack-years and whether initiating screening at age <55 years would result in additional lung cancer deaths being averted [ref.]."

With these premises, the authors set out to determine "the annual number of lung cancer deaths that could potentially be averted . . . assuming the screening regimen adopted in the NLST is fully implemented among the entire screening-eligible population in the United States."

To this end, they derived estimates of the numbers of people in the U.S. in 2010 who satisfied the admissibility criteria of the NLST, numbers specific to gender and age and whether current or former smoker, supplementing these by estimates of the corresponding rates (incidence densities) of death from lung cancer. As an example of the calculations, for men 55–59 years of age, the total population size was said to be 9,523,648; in it, the prevalence of eligible current smokers was estimated to be 9.91 %, and their rate of death from lung cancer was estimated to be 381.0/10^5y. From these inputs (with striking numbers of significant digits), the number of deaths from lung cancer per 1 year was calculated as 9,523,648 (0.0991)(381.0/10^5y) = 719/y. Correspondingly, the deaths of former smokers were calculated to be 421/y, implying a total of 1,140/y for this stratum by gender and age. Summation over all of these strata gave 12,250/y.

The authors' Conclusion (*sic*) was that "The data from the current study indicate that [low-dose CT] screening could potentially [*sic*] avert approximately 12,000 lung cancer deaths per year in the United States." In the Editorial associated with this report, L. Kessler's opening was this: "Ma et al. have estimated that if every one of the current or former smokers in the United States who has accumulated 30 pack-years began low-dose screening with computed tomography (CT), each year approximately 12,000 deaths due to lung cancer could be delayed [*sic*] or averted [ref.]. These estimates are based on the highly successful results of the [NLST] published last year [ref.]." In the text that follows, the section headings continue the laudation of the NLST: "The remarkableness of this advance in attacking lung cancer"; "Why are these estimates important?", . . .; "The 20% loaf and a call for national policy?"

Now, insofar as these estimates indeed are important, this actually is because they are a dramatic example of the *misguidedness of intramural scientific 'experts' in agencies such as the ACS* – agencies with a major role in sponsoring what they regard as policy-relevant health research, propensity to use the results of such research to promulgate 'evidence-based' public-policy recommendations in regard to practice of healthcare, and control of a journal to do just this.

Contrary to its title, etc., the truth is that this ACS "study" did not address any "annual" number of preventable deaths from lung cancer: The background of this "study" was the *NLST result* that, in the CT arm, 356 deaths from lung cancer were 'observed' while 445 were 'expected' – indicating a *reduction of 89 deaths* (89 being 20.0 % of 445). And the actual aim of the ACS calculations (above) seemingly was to *derive the equivalent of this NLST result* for the hypothetical situation in which a cohort would have been formed of all of the 8.61 million Americans who in 2010 satisfied the eligibility criteria of the NLST.

Those calculations were seriously at variance with this aim. In the NLST, the 'expected' number (445) of deaths from lung cancer in the CT screened accrued over the duration of follow-up of the cohort so screened; it was not the number that 26,722 persons such as were entered into the trial's CT arm would suffer in 1 year. By analogy, then, the 'expected' number of deaths from lung cancer in the cohort of 8.61 million members would not be properly represented by the estimated 61,250 deaths from lung cancer in the eligible population in $2010 - 61,250(0.200) = 12,250$.

The stratum-specific 'expected' numbers should have been derived on the basis of their counterparts in the CXR arm of the NLST and added up, thus obtaining a number much larger than the 61,250 (as the counterpart of the 445). For, the idea in the calculations really was hypothetical replication of the NLST for a cohort that is different in size and also in its distribution by the stratification factors. On the premise of identical structure of the hypothetical population to the one entering the NLST, the result would have been, simply, $8.61 \times 10^6/320 = 26,900$, given that, in the NLST, "The number needed to screen with low-dose CT to prevent one death from lung cancer was 320" (ref. 7).

The correctly derived, much larger number *would still have been an underestimate* of the number of deaths prevented by the 'screening' (by its associated early treatments in place of late ones), because not all of the deaths prevented in the NLST's CT arm would have otherwise occurred during that trial's follow-up.

But the number at issue here is an odd one: it presupposes practice in which the 'eligible' persons are screened à la NLST: for up to three annual rounds, once in a lifetime. *That number has nothing to do with annual screening of the population at issue here.* For, that population is not a cohort formed from the 'eligible' population but that population as it is – the dynamic, ever new population of some eight million Americans 'eligible' to be screened, annually or with whatever frequency. If this population were annually screened, there would be an annual number of attained cures of otherwise fatal cases; and there thus would be their consequent *annual reductions in future deaths from lung cancer.*

This being the last example study in this series of critical examinations of recent studies to advance the knowledge-base of medicine, it prompts these closing observations specific to intervention-prognostic research:

1. In countries such as the U.S., the greatest health concern of people is the possibility of coming down with a case of cancer; and by the same token, their greatest hope in respect to healthcare is that, in the event of cancer getting to be diagnosed, it would be cured by what modern medicine can offer.
2. In respect to the most common types of cancer, lung and breast cancers among these, the hopes for cure now center on 'screening' for the cancer, meaning pursuit of the cancer's early, latent-stage detection and its consequent early treatment.
3. The NLST epitomizes the response to these hopes on the part of the preeminent sponsors of cancer research in the U.S., namely the NCI and the ACS: They sponsored a very major *study to test the hypothesis* (*sic*) that 'screening' for lung cancer, with low-dose CT imaging the first test in the diagnostic algorithm, serves its purpose – of providing for curative, life-saving treatment of otherwise incurable, fatal cases of the cancer. And they were content to sponsor a *study that did not provide for quantitative assessment* of how commonly the early treatment, provided for by diagnosis (rule-in) under the screening, is curative in otherwise incurable cases; it did not matter to them that the study was not designed to quantify the gain in the cancer's curability rate – the reduction in its case-fatality rate – in the (likely) event that the screening indeed turns out to serve its purpose.
4. While the NLST thus was hugely wasteful of the opportunity to meaningfully quantify the 'screening's' intended consequence (as outlined in the context of study 5), this problem has been seriously compounded by opinion-leaders' publicly-expressed beliefs that it did what it should have done. Thus, eminent medical societies are 'guidelining' clinicians to tell their clients, concerned to learn about the utility of CT screening for lung cancer, (the falsehood) that it (per the NLST) serves to provide for cure in 20 % of otherwise fatal cases (study 5 above). And one of the two agencies sponsoring the NLST is telling epidemiologists (the falsehood) that their interest in the attainable reduction in the annual number of deaths from lung cancer in the screening-eligible population was addressed by the NLST (study 6 here).

5. This very eminent topic – potential for cure of lung cancer by screening-provided opportunity for early treatment – brings to focus an important question: Why is there no progress in the design – objects design and methods design – of these studies, and even eminent misrepresentation of their results? While science supposedly is self-correcting, why are the necessary corrections not taking place in this context? Why are there only exhortations of the high standards of the requisite science that purportedly is represented by studies such as the NLST? Why practically no voices of dissent?

I suggest two answers to that important question. First, as the intervention-prognostic topic is intertwined with a diagnostic one, and concerns of community medicine (re mortality rates) are intertwined with those of clinical medicine (re curability rate or case-fatality rate), understanding of the issues of study design and quantitative meaning of the study results is *exceptionally challenging*. And second, progress in these matters (like others in science) would presuppose public critique of the status quo; but this is stifled when agencies such as *the NCI and the ACS assume too many roles*. Not only are they the principal sources of funding for cancer research but also propagators of orthodoxy in respect to the directions, principles, and significance of the research. In this context, heterodoxy is typically kept *sub rosa*.

Chapter 8
Derivative Research for Scientific Gnosis

Contents

8.0 Abstract

In derivative research for scientific gnosis, the objects of study are those of original studies for it. In derivative research, however, the evidence that is produced is not original but derived by synthesis of contributions from select ones of the original-study results on the gnostic (dia-, etio-, or prognostic) probability function in question.

While coping with publication bias in the aggregate of published reports from original studies is a very major challenge in derivative studies of the genre of hypothesis-testing (qualitative), consequent to the results from those original studies commonly being classified as 'positive' or 'negative' and the preferential publication of 'positive' results, this problem does not extend to derivative research for the advancement of the knowledge-base of scientific gnosis. For the result, whether original or derivative, on a gnostic probability function, GPF is not classifiable as 'positive' or 'negative.'

The principal challenge in derivative research on a GPF is, at present, proximal to any concern about validity of it. The principal challenge still is finding any (sic) original study on a meaningfully conceived GPF on the gnosis at issue.

O. S. Miettinen, *Toward Scientific Medicine*, DOI 10.1007/978-3-319-01671-9_8,
© Springer International Publishing Switzerland 2014

8.1   The Essence of a Derivative Study for Scientific Gnosis

Medicine is scientific insofar as its theoretical framework is rational and its knowledge-base derives from science (Preface, Sect. 4.1). The rationality and knowledge-base at issue here have to do with the very core of medicine: the way of thinking about the client's health and about the type of general medical knowledge that is needed for attainment of esoteric knowing – gnosis (dia-, etio-, or prognosis) – about this, and commitment to deployment of that type of knowledge in the pursuit and attainment of gnosis.

In scientific medicine, gnosis is thought of in terms of gnostic probabilities, and the knowledge-base of pursuing and attaining gnosis is taken to be gnostic probability functions from gnostic science. The production of a GPF is, most broadly, tri-phasic: the first phase is the design of its form; the second phase is research to produce evidence of the designed form; and the third phase is transmutation of the evidence about the GPF into knowledge about it. The research is bi-phasic: original research is followed by derivative research based on this.

The definitional essence of a *derivative gnostic study* is: a project to produce a synthesis of pre-existing original evidence on a particular GPF for a particular domain, the derivative evidence consisting of the obtained synthetic counterpart of the original results on the GPF for its domain, this together with documentation of the genesis of this synthetic result, the way it was derived from the existing aggregate of original evidence on the GPF at issue for the domain at issue. There can be more than just one derivative study on the same aggregate of existing original evidence on the GPF at issue. If they deploy, faithfully, the same protocol, they produce the same result. (This contrasts with original studies with faithful application of a protocol that is the same in all validity-relevant respects.)

A derivative gnostic study for scientific medicine involves, in respect to an objectively defined GPF for an objectively defined domain, these elements: identification of all original studies on that GPF for that domain; selection of some, or all, of them as providing inputs to the derivative study; and then synthesizing either the data or the results of these studies into the derivative – synthetic – result and documenting the genesis of this result in those three respects.

The evidence from a derivative gnostic study for scientific gnosis is analogous to that from an original study for that purpose: the result for a given GPF for is domain, together with the genesis of this. The difference is only that a derivative result is more precise than any of the original results from which it is derived.

8.2   The Validity of a Derivative Study for Scientific Gnosis

Derivative research for scientific gnosis – inherently quantitative (as to the magnitudes of the parameters in a GPF) – is, commonly, preceded by hypothesis-testing – inherently qualitative (as to some potential component in a GPF, its relevance for involvement in that GPF).

The evidence from an original qualitative study is commonly classified as 'positive' or 'negative,' according as it is seen to support the hypothesis or to detract from it. And whereas there commonly is a greater concern among investigators to submit the report on 'positive' evidence for publication, and also a greater willingness among journal editors to accept 'positive' evidence for publication, published evidence from qualitative studies is, generally, marred by quite serious publication bias, and this represents a major challenge for derivative research of the hypothesis-testing type. The meaning of this 'publication bias' is not that the published studies are prone to be biased. Instead, the meaning is that the published original studies constitute a biased segment of all of the studies that have tested the hypothesis, due to selective absence of studies with 'negative' results.

The *evidence from an original study on a GPF* is not, ever, classified as 'positive' or 'negative,' and derivative studies on GPFs, based on published evidence on them, therefore are *not prone to be publication-biased* the way their counterpart for hypothesis-testing are.

While publication bias is the principal challenge, by far, for the validity of derivative studies of the genre of gnosis-related hypothesis-testing, in derivative research on actual GPFs there is, at present, a challenge much greater than that of validity assurance. Synthetic evidence, however precise when derived from its less precise inputs, is meaningful for the advancement of the knowledge-base of scientific gnosis only if it addresses, validly, a meaningfully defined GPF for a meaningfully defined domain of it – by virtue of these requirements of meaning-assurance having been shared by the original studies involved in the synthesis.

When a derivative study for scientific gnosis addresses a meaningfully-defined GPF for a meaningfully-defined domain for this on the basis of valid original evidence on this function, there are no notable challenges for the attainment of validity for the derivative evidence. As it is a present, the principal challenge, by far, is in finding any (sic) inputs to such a meaningful derivative gnostic study (Sect. 8.3 below).

A derivative study is *not properly referred to as a 'systematic review'*; a better term indeed is 'derivative study' (of a GPF). For a derivative study naturally must be systematic – "methodical, done or conceived according to a plan or system" (my OED) – just as any original study must be; and review of available original evidence inherently is involved. And for what is done with the original data/results involved in a derivative study, *'synthesis' rather than 'meta-analysis' is the proper term* – synthesis and analysis being, respectively, "the process or result of building up separate elements" and "a detailed examination of the elements or structure of a substance etc." (my OED).

8.3   Some Recent Studies Critically Examined

Regarding *diagnostic* research, the 2012 volumes of *JAMA* reported on several 'systematic reviews' under "The rational clinical examination." The first one of these was:

> *Reference 1:* Udell JA, Wang CS, Timmouth J et al (2012) Does this patient with liver disease have cirrhosis? JAMA 307:832–842

The Objective had been "To identify simple clinical indicators that can exclude or detect cirrhosis in adults with known or suspected liver disease." To this end the authors "searched MEDLINE and EMBASE and . . ." They "retained 86 studies of adequate quality that evaluated the accuracy of clinical findings for identifying histologically proven cirrhosis."

The authors did not address any (pre-designed) probability function for the presence of cirrhosis in an objectively defined domain of patient presentation, synthesizing the results of valid studies on this.

Next,

> *Reference 2:* Srygley FD, Gerardo CJ, Tran T, Fisher DA (2012) Does this patient have severe gastrointestinal bleed? JAMA 307:1072–1079

In this study the Objectives had been "To identify the historical features, symptoms, signs, bedside maneuvers and basic laboratory results that distinguish acute upper GIB (UGIB) from acute lower GIB (LGIB) and to risk stratify those patients with UGIB least likely to have severe bleeding that necessitates an urgent intervention." To this end the authors conducted "A structured search of MEDLINE (1966-September 2011) . . ." and "High-quality studies were included . . ." Then, "One author abstracted the data (prevalence, sensitivity, specificity, and likelihood ratios [LRs]) and assessed methodological quality, with confirmation by another author. Data were combined using random effects measures."

The authors did not address any (pre-designed) probability function for the presence of gastrointestinal bleeding, synthesizing the results of valid studies on this. Their Conclusions were about which findings "increase the likelihood of UGIB" and which ones "make UGIB less likely." Besides, "Blatchford clinical prediction [sic] score ... is very efficient [sic] for identifying patients who do not require urgent intervention."

Then the third one in this series:

Reference 3: Nishijima DK, Simel DL, Wisner DH, Holmes JF (2012) Does this adult patient have a blunt intra-abdominal injury? JAMA 307:1517–1527.

Its Objective had been "To systematically assess the precision and accuracy [sic] of symptoms, signs, laboratory tests, and bedside imaging studies to identify intra-abdominal injuries in patients with blunt abdominal trauma." To this end the authors conducted "a structured search of MEDLINE ... and EMBASE ..." They "included studies of diagnostic accuracy for intra-abdominal injury that compared at least 1 finding with a reference standard of ...

Again, no synthesis of results on a given type of diagnostic probability function for an objectively defined presentation domain was produced.

Finally, this fourth one:

Reference 4: Coburn B, Morris AM, Tomlinson G, Detsky AS (2012) Does this adult patient with suspected bacteremia require blood cultures? JAMA 308:502–511

The Objective had been "To review the accuracy of easily obtained clinical and laboratory findings to inform the decision to obtain blood cultures in suspected bacteremia." So, this was yet another "review" of original studies for diagnosis without concern – and opportunity – to synthesize the results of various valid original studies on the same diagnostic probability function.

As for 'systematic reviews' of studies for etiognosis or prognosis in the 2012 volumes of JAMA, quite analogously with diagnostic research, there was none synthesizing evidence from original studies on a particular rate ratio function or prognostic probability function. Clearly, research for the knowledge-base medicine has not yet come to adopt the view that this research needs to address functions, gnostic probability functions.

Chapter 9
From Gnostic Research to Gnostic Knowledge

Contents

9.0 Abstract

Even though the research for the knowledge-base of genuinely scientific medicine (Sect. 4.1) remains essentially non-existent, medicine should already be quasi-scientific – rational in its theoretical framework and driven by gnostic probability functions in expert systems, with those GPFs based on non-scientific expertise. That expertise should be cultivated by extensive reviews of documented cases, starting in the discipline's training program already, with personal experience only a minor supplement to this; and it should be garnered, ultimately in well-designed GPF form, from experts' gnoses in hypothetical cases.

Once evidence from research on a well-designed GPF becomes available, its transmutation into scientific knowledge about it becomes possible upon the requisite innovation of medical-scientific journalism.

Both the quasi-scientific and the scientific GPFs define normative gnostic practice, but they do not imply justifiable guidelines for decisions in practice.

O. S. Miettinen, *Toward Scientific Medicine*, DOI 10.1007/978-3-319-01671-9_9, 171
© Springer International Publishing Switzerland 2014

9.1 From Experts' Gnoses to Gnostic Functions

Now that we already are several decades into this Information Age, the knowledge-base of medicine should already have been codified in the form of gnostic (dia-, etio-, and prognostic) probability functions (Chap. 2) – accessible, as needed for gnostic probability-setting, from suitably up-to-date gnostic expert systems (Sect. 4.8). But this knowledge-codification has not yet taken place and, consequently, the needed expert systems do not yet exist. So medicine still goes by mere subjective opinions about gnostic probabilities in lieu of knowledge about them. And, unsurprisingly, those case-specific probability opinions are quite strongly doctor-specific even among the most expertly of doctors on the types of case at issue.

In the transition from this very unsatisfactory state of affairs to the glory state of scientific medicine, a way station would be *quasi-scientific* medicine. In this, as in scientific medicine, the theoretical framework would be rational, and gnostic probabilities would be set by consulting of the best available gnostic probability functions. But these GPFs of quasi-scientific medicine would not express scientific knowledge about the probabilities they address. Instead, they would give *experts' typical probabilities* (still profile-conditional) as ersatz values for these, ones without basis in research on these GPFs.

Any practitioner, even an expert on the gnosis at issue, should be concerned to know what experts' typical value is for the probability in question, and to adopt this as their own best understanding. And in any case, this is what the doctor's client in their particular case is concerned to learn (rather than the doctor's personal opinion).

This raises the question: How could those quasi-scientific GPFs be constructed? This is a question of methods design, with objects design unchanged from what it is in the context of gnostic research for truly scientific medicine (Chaps. 5, 6 and 7).

A GPF of a given form for a given domain commonly implies an enormous number of probabilities each of which can be of concern in practice, each specific to a particular gnostic profile implied by the set of realizations of the statistical variates (Xs) of the model in the case in question from the function's domain. For the development of a quasi-scientific GPF, a number of these profiles – at least a few scores of those case vignettes – need to be selected as hypothetical cases for the members of a select expert panel to consider. Each of the members specifies, independently of the others, what their best understanding is of the correct gnostic probability in each of those cases, and the typical value – median – of these probabilities is identified for each of the cases. Given the thus-obtained data, a General Linear Model is fitted to them, with the dependent variate the logit of the median probability and the independent variates the ones in the designed logistic model for the probability.

Rather than further particulars of this codification of the tacitly prevailing, still pre-scientific (*sic*), knowledge-base of medicine, some comments on the concept of *expert* physician in a principally knowledge-based discipline of medicine are in order, given the central role of expert gnosticians in this process. This is so for the added reason that medical dictionaries – my Dorland's and Stedman's at least – do not give the medical meaning of 'expert.' "Medical expert/clinical decision maker" is, however, being presented as one – the very first one – of the six "essential competencies for Canadian specialist physicians"; but, remarkably, the stated particulars of this are not at all about expert level of knowledge relevant to practice, nor even about expert quality of practice-relevant opinions (Sect. 3.4).

My OED says that generally meant by 'expert' can be "a person who has a comprehensive and authoritative knowledge or skill in a particular area," or else meant is this quality of a person. But when the particular area is the domain of a GPF, no doctor now has a comprehensive and authoritative knowledge about the gnostic probabilities at issue in this domain; such expertise does not exist in today's medicine, pre-scientific – specifically, pseudo-scientific (Sect. 4.1) – as it remains (Chap. 3).

As some of the rationally-defined disciplines of medicine would be thought of as being knowledge-based (rather than skills-based; Sect. 4.2), this they would be in principle, though not in present-day reality. In reality we still have only *opinion-based medicine*, OBM, in lieu of, even, the quasi-scientific precursor of the scientific version of knowledge-based medicine, KBM.

But the opinions (about gnostic probabilities) in OBM are not to be viewed in relativistic terms, as though they all were to be weighted equally when the concern is to surmise the level of the correct probability (objective; Chap. 1). Some opinions are to be regarded as being more expert than some others. Varying *degrees of expertise* in the context of a particular category of gnostic challenges thus is, definitely, an ingrained and sound concept in medicine. The question only is the particular nature and identifiability of this quality in a given type of context.

An opinion as an ersatz for knowledge in OBM is, I suggest, expert – and by the same token, the opiner of the probability (profile-specific) at issue is an expert on it – on the ground of the opiner's superior *basis for the opinion*. In today's medicine (pre-scientific), superior basis for a gnostic opinion is commonly and justifiably taken to be that of having most experience with cases from the domain of gnostic challenges of which the case at issue is an instance. Meant by 'experience' in this is personal, direct experience – with cases from the presentation domain of 'adult with acute chest pain,' for example.

The relevant experience can, however, be indirect; and as practice experience is valued in the absence of evidence from directly practice-relevant (gnostic) research, preparation for practice in one of the disciplines of medicine that in principle are knowledge-based should involve, centrally, *review of a large number of cases* from that discipline's domain (in lieu of what now is involved; Sect. 3.4). This would prepare the trainee's mind to surmise the probabilities (correct) in yet other profile-specified cases from the domain – within the inherent limitations of this case-based inductive learning.

Such are now the limitations of experience-based expertise that doctors with the same extensive experience tend to have highly divergent opinions about the probabilities in particular profile-specific cases with which they are confronted, whether actual cases in practice or hypothetical cases presented to them as members of an expert panel for the development of an expertise-based GPF. But as I noted above, the statistical-type theoretical framework sketched above allows formal extrapolation from experts' typical probability-opinions in a particular set of cases from the domain at issue to the counterparts of these in yet other cases from this domain; and the point about preparation for practice in a discipline of knowledge-based medicine above points the way to cultivating outstandingly high expertise for exploitation in the development of the knowledge-base of quasi-scientific medicine, this within the inherent limitations of the case-based method of learning about gnostic probabilities.

9.2   From Gnostic Research to Gnostic Knowledge

Once valid and reasonably precise evidence from research on a (well-designed) GPF gets to be available, the probabilities it defines are not matters of personal opinion; they are fundamentally different from these, matters of science rather than pre-scientific personal opinion. But they are not inherently matters of scientific knowledge.

Scientific knowledge, too, is a matter of opinion: when a GPF represents scientific knowledge, it represents the relevant *scientists' shared opinion*, first about the way to think about the magnitude of the probability in the domain at issue, expressed by the form of the function; and it also represents their shared opinion about the correct magnitudes of the parameters involved in the designed GPF.

Those shared scientific opinions also are ones of *experts*, now experts on the type of research that underlies the knowledge. The relevant scientists, collectively, master the design of the GPF for the domain of this all the way to the functions's statistical form, akin to the clinical experts together with an objects-design expert in the development of a quasi-scientific GPF (Sect. 9.1 above). But their added expertise is in the design and evaluation of the methods of study for the genre of gnosis at issue.

In the appropriate culture of medical-scientific journalism (Sect. 4.5), once this gets to prevail, the available evidence on the GPF in question will be critically discussed in the relevant journals, by members of the relevant scientific community. If and when this discussion leads to a shared opinion about the evidentiary issues involved, discussion of the inferential issues ensues. Insofar as this results in shared opinions – quite tentative perhaps – about the magnitudes of the parameters involved, scientific knowledge about the GPF has been achieved.

9.3 Focusing Guidelines on Gnostic Knowledge

Once there is a GPF fairly representing the typical probabilities of a panel of genuine experts (defined in Sect. 9.2 above) or the shared beliefs about the magnitudes of its parameters in the relevant scientific community, this GPF, even if only quasi-scientific, defines normative practice in its domain for setting the gnostic probabilities it addresses. It represents guidelines for practice in this segment of knowledge-dependent practice – as for what a doctor is to take the relevant gnostic knowledge to be.

A GPF does not imply rational guidelines for decisions in practice, not about the submission of a patient to a particular diagnostic test (coronary angiography, say) or a particular intervention (coronary artery bypass surgery, say). For important added inputs to such decisions are the valuations of the consequences of the decisions/ actions being considered, and these are not part of the knowledge-base of medical practice; they are specific to, and knowable only by, the patient.

The appropriate decision-maker thus is not the doctor but the patient, and development of practice guidelines in reference to doctor's decisions – now commonplace even in the absence relevant and appropriate GPFs – reflects misunderstanding of what knowledge-based medicine, quasi-scientific or scientific, fundamentally is about.

Epilogue

As may be recalled from Sect. 3.4, the Faculty of Medicine of my McGill University endeavors, in its undergraduate education already, "to ensure career-long excellence in whole-person care"; and in this, a "fundamental premise" of the faculty is that: "The basic sciences and scientific methodology are fundamental pillars of medical knowledge."

According to the present oeuvre, fundamental to the development of practice-guiding medical knowledge should be seen to be recognition and appreciation of *the inherent nature, great complexity, and necessary form of codification of the requisite knowledge-base of medicine*, all of which I expounded in the opening two chapters. Accordingly, while the chapter that immediately followed was an essay on the present, highly wanting state of practice-guiding medical knowledge (Chap. 3), in the main the ensuing chapters were about the development of genuinely knowledge-based and ultimately genuinely scientific medicine (Chaps. 4, 5, 6, 7, 8, and 9).

Looking back at all of this, I now condense into a mere quadruplet of theses my conception of the necessary reformations of the knowledge culture of medicine – well fewer than Luther's 97 theses in Wittenberg in 1517, at the launching of his guest for reformation of the culture of his concern at his time.

1. The knowledge-base of each of the knowledge-dependent disciplines of medicine – disciplines to be defined by their respective coherent and self-contained aggregates of the requisite knowledge (Sect. 4.2) – needs to be codified in terms of *gnostic probability functions* (Chap. 2). For without these GPFs and their as-needed accessibility from practice-guiding *expert systems* (Sect. 4.8), practice in the knowledge-dependent disciplines of medicine will generally remain only opinion-based rather than in any meaningful sense knowledge-based (Sect. 9.1), even in this Information Age.

2. The development of those GPFs needs to be *quasi-scientific* before it can be scientific (Sect. 9.1). And preparatory to this, availability of pre-scientific gnostic experts needs to be cultivated through replacement of the prevailing type of

O. S. Miettinen, *Toward Scientific Medicine*, DOI 10.1007/978-3-319-01671-9,
© Springer International Publishing Switzerland 2014

'specialty training' (Sect. 3.4) by early accumulation of great experience – second-hand, from reviews of large numbers of case records from one's particular discipline of knowledge-dependent medicine (Sect. 9.1).

3. Development of GPFs for genuinely *scientific* medicine (Sect. 4.1) requires an even more radical reformation of the knowledge culture of medicine, as GPFs are not yet being addressed in medical textbooks (Sects. 3.1 and 3.2) nor in medical journals (Sect. 3.3), and the research culture behind all of this remains rife with notable aberrations of principles even apart from its principal flaw of not even addressing GPFs (Part III).

4. Parallel with those developments being put in motion, *education* for the knowledge-dependent disciplines of medicine needs to be weaned away from its prevailing "fundamental premise" (above) and to undergo radical restructuring. The urgent need is, first, to remove the bulk of the existing content of the undergraduate education, recognizing how irrelevant it actually is for any suitably-defined discipline of modern medicine (Sect. 4.2) and, hence, how wasteful it is of time and other resources; and second, to introduce, to both under- and postgraduate education, content that truly is fundamentally relevant yet missing at present (Sects. 4.6, 4.7, and 9.1). A sharp distinction is to be made between the education of *practitioners* (Sect. 4.6) and that of *researchers* concerned to help advance the knowledge-base of practice (Sect. 4.7).

Implementation of these reforms, radical as they are, would require a worldwide commitment to it by leaders of medicine and a globally co-ordinated process. It thus likely would require leadership by the World Health Organization.

Glossary

Diagnosis/Etiognosis/Prognosis See Gnosis

Disease/Defect/Injury See Illness

Etiogenesis The genesis of an illness-defining anomaly as a matter of the causation of its pathogenesis, or the causal origin of a sickness not manifesting illness. Cf. Pathogenesis

Gnosis (in medicine) A physician's esoteric knowing – probabilistic – about a client's health. See Knowing; Knowledge

    Diagnosis Gnosis about a particular illness as to its presence/absence in the client at present or at some time in the past

    Etiognosis Gnosis about an antecedent/concomitant of the clien'ts illness/sickness as to its involvement/non-involvement in the etiogenesis (causal origin, etiology) of the illness or sickness not manifesting illness

    Prognosis Gnosis about the future course of a client's health as to the occurrence/non-occurrence of a particular phenomenon of this

Illness (syn. 'ill-health'; ant. 'health') Somatic anomaly at the root of sickness, potentially at least (e.g., trisomy 21; cancer)

    Defect (L. *vitium*) Illness in which the defining somatic anomaly is a stationary state (e.g., trisomy 21; neural tube defect; phenylketonuria; myocardial fibrosis)

    Disease (L. *morbus*) Illness in which the defining somatic anomaly is a process resulting from its precursor process (somatic) of pathogenesis (e.g., cancer; mycobacterial tuberculosis)

    Injury (L. *trauma*) Illness in which the defining somatic anomaly is a process resulting from a destructive stress on the tissue(s) involved (e.g., wound; fracture; burn)

Knowing Awareness of knowledge (e.g., about a gnostic probability)

Knowledge Experts' intersubjective, shared perception of truth (which may not be true)

    Empirical k. (syn. 'substantive' k.) Knowledge in which the perceptions of truth are based on experience together with reasoning

    Theoretical k. Knowledge in which the perceptions of truth are based on reasoning alone (Kantian 'pure reason')

O. S. Miettinen, *Toward Scientific Medicine*, DOI 10.1007/978-3-319-01671-9,
© Springer International Publishing Switzerland 2014

Medical science Science intended to advance medicine (practice of it); that is, medicine-oriented 'applied' science

'Basic' m. s. Science intended to result (through development) in novel products for use in medicine (gnostic tests; interventive agents)

Quintessentially applied m. s. Science intended to advance the knowledge-base of medicine; that is, research – original and derivative – producing evidence, and the induction from this – by the relevant scientific community – of knowledge about the object (gnostic) at issue

Medicine Physicians' professional practice of healthcare

Clinical m. Medicine serving individuals, one at a time. Cf. Community m

Community m. (syn. 'epidemiology') Medicine serving a population, as a Population (rather than individually). Cf. Clinical m

Evidence-based m. Medicine in which the physician uses evidence-informed personal opinions as ersatz knowledge, in setting gnostic probabilities. Cf. Scientific m

Knowledge-based m. Medicine in which the physician deploys (not personal opinions but intersubjective) knowledge, theoretical and substantive, in setting gnostic probabilities. Cf. Scientific m

Scientific m. Knowledge-based m. in which the physician deploys a rational theoretical framework – rational theoretical knowledge, that is – and the requisite substantive knowledge from science, in setting gnostic probabilities

Thought-based m. Medicine (Flexnerian) in which the physician deploys scientific-type thinking, learned in laboratories of 'basic' medical sciences, in solving problems of patient care (diagnostic ones, most notably). Gnostic probabilities are not among the objects of this thinking. Cf. Scientific m

Pathogenesis The genesis of an illness-defining somatic anomaly as a matter of the changes from normal tissue to that anomaly. Cf. Etiogenesis

Patient Physician's client who is sick

Physician Graduate of medical school, licensed to practice medicine and actually practicing it

Science See Medical science

Scientific Having to do with science or being an application of science

Sickness Somatic state from which a person directly suffers (e.g., nausea, Down's syndrome). It may not be a manifestation of an illness but merely of a stressful circumstance (e.g., being in a moving car or boat; using a medication)

Index

O. S. Miettinen, *Toward Scientific Medicine*, DOI 10.1007/978-3-319-01671-9,
© Springer International Publishing Switzerland 2014

Printed in the United States
By Bookmasters